A HISTORY OF
GWR GOODS WAGONS

Volume 2
WAGON TYPES IN DETAIL

A. G. ATKINS, W. BEARD,
D. J. HYDE and R. TOURRET

DAVID & CHARLES
NEWTON ABBOT LONDON
NORTH POMFRET (VT) VANCOUVER

ISBN 0 7153 72904
Library of Congress Catalog Card Number 76–45508

Set in 10 on 11pt Plantin
and printed in Great Britain
by Redwood Burn Ltd Trowbridge
for David & Charles (Publishers) Limited
Brunel House Newton Abbot Devon

Published in the United States of America
by David & Charles Inc
North Pomfret Vermont 05053 USA

Published in Canada
by Douglas David & Charles Limited
1875 Welch Street North Vancouver BC

CONTENTS

*S6/7 type poultry and fish truck with barred sides,
photographed at Churston about 1906. Planked ends
and fall-down doors, as opposed to swing sideways
doors. Body mounted on ex-broad gauge four-wheel
carriage underframes. Some barred fish trucks reputed
to have had cabooses. 4–4–0ST No 13 rebuilt 1897
from 2–4–2T, on loan to Liskeard and Looe Railway
between 1901–8.* (Courtesy Photomatic Collection)

ABBREVIATIONS

BG	Broad Gauge
BGW	'British Goods Wagons' by R. J. Essery, D. P. Rowland & W. O. Steel
cf	Compare with
DC	Dean-Churchward brakes (Marks I, II and III)
GWR GW	Great Western Railway
GWS	Great Western Society
L	Lot Number
NCU	Not Common User
NWO	New Work Order
oh	Over headstocks
OK	Proprietary type of oil axlebox
os osL	Old series of lot numbers
PO	Private owner
qv	which see
RC	'A Pictorial Record of GWR Coaches, Part 1' by J. H. Russell
RCH	Railway Clearing House
RCW	Railway Carriage & Wagon Company (Birmingham and Gloucester)
RW	'A Pictorial Record of Great Western Wagons' by J. H. Russell
RWA	'Great Western Wagons Appendix' by J. H. Russell
s/c	Self-contained
SWB	Special Wagon Book
SWR	South Wales Railway
u/f	underframe
wb	wheelbase
WD	War Department

PREFACE

THIS is the second of two books which present a historical record of the development of goods rolling stock on the Great Western Railway. The whole family of designs is inter-related, reflecting changes in constructional details and operational needs. They present a picture of the GWR freight scene from the 1870s to nationalisation.

Volume 2 works through the wagon index, discussing each wagon design in detail and showing how it changed. Drawings are given in 4mm/ft scale such that 'identikits' can be assembled of some 450 different designs. Volume 1, it will be recalled, dealt with GWR freight working, GWR telegraph code names, the index of wagon diagrams and the general construction of wagons.

The overall documentation is the result of extensive independent research at Swindon and elsewhere, and the books represent years of collation of information and careful sifting of data to reconcile contradictory stories to establish, for the first time in many cases, the details of the lives of GWR wagons. Some published information has been incorrect, but it would be foolish to imagine that there are no errors in our work and we should appreciate hearing about them.

Thanks are owed to many officers of British Rail, particularly Messrs Day, Frankling, Tanner, Hartley and their colleagues in the Swindon drawing office, and Messrs Chilvers and Froud of the Swindon photographic studio, for considerable help and guidance extending over many years. Most of the photographs presented here are from the magnificent GWR collection and are reproduced by courtesy of British Rail. Likewise the drawings are based on Swindon diagrams.

Where we have drawn upon the work of others, we have given acknowledgement and referenced the source material. Mrs Margaret Woodfin was most kind in giving us access to her late husband's records (much of which is contained in the Historical Model Railway Society archives), while in the preparation of the manuscript Mrs Meg Atkins patiently bore the brunt of the typing, for which we are immeasurably grateful.

We should like to thank people who have written about Vol. 1, and add to our acknowledgements list the names of Messrs D. M. Lee (who painstakingly cross-checked the diagram index, Table 7, Chapter 3), W. Hillier, J. Hodgetts of GWS, R. Metcalf, R. Miller, J. Slinn, Guy Williams and Pendon Museum, and K. Williams. Many of their comments, along with our own emendations and clarifications, are incorporated in this volume.

Among the obvious errors in Vol.1 are the transposition of the photograph legends on pp 49/50 (i.e. O32 is on p.49 and vice-versa); the omission of V37/8 in the index of wagon diagrams; the serial numbers given for J1 in the index; the dates of the MICA A/B codes in Chapter 2. Details of these and other corrections (e.g. bogie loco coal numbering) are given in this volume in the relevant sections. Finally, a wrong impression given in Vol.1 by the use of 'brakeblocks swung from the solebar' should be corrected. Brakeblocks were not swung from the solebar as such: what was intended to be conveyed was that this type of brakeblock, dating from the last quarter of the 19th century, had no pin in the linkage at the point of attachment to the block, and thus swung 'rigidly' from its axle up *inside* the chassis.

A. G. Atkins, W. Beard, D. J. Hyde and R. Tourret

Above : 36951 in broad gauge storage sidings, Swindon, said to be 1898. Standard gauge C8 CROCODILE C (old code) 36951 in original type II guise at centre. Wooden deck to well, open end platforms, screw brake, grease boxes. On left, A3 POLLEN in original form. Bolsters, not turntables. Note 32994 marshalled next to 32996. Fish wagons (later S4/5) in foreground, various BG vehicles behind.

Below: V30 ALE wagon, photographed in 1939 when converted from W1 MEX built 1895. One sided lever brake, toothed rack, still has old brake blocks rigidly connected to swing arms. Wright-Marillier partition device still in place. Screw couplings, 1ft 8½in buffers. Cross bracing on ends of W1 as built, replaced by angle stays (see p.99).

CHAPTER 1
A – ARTICULATED WAGONS FOR BOILERS, GUNS & BRIDGE GIRDERS
(Pollens)

First place in the diagram index was allotted to iron girder wagons intended for long heavy loads. In pairs, or groups of four, these wagons when first introduced in the 1880s were longer than other heavy wagons such as CROCODILES and BEAVERS. On occasion a POLLEN set could be split, and the component wagons put at the ends of very long structures.

Just before the conversion to standard gauge, the GWR possessed three sets of these wagons. A group of four 4-wheel 12-ton wagons (59ft 3in over headstocks) had been built in 1885 under osL331, numbered 32993-6, and later coded POLLEN (see photograph in Chapter 3 following). Another group of four 4-wheel wagons (86ft over headstocks) had been built between 1887 and 1890 under osL418/528, numbered 32989-92, and later coded POLLEN A; each 'outside' wagon could take 20 tons whereas the centre wagons were 12-tonners, giving a capacity of 64 tons. The third set was a pair of broad-gauge 6-wheel 30-ton wagons (51ft over headstocks), built 'convertibly' under osL422 in 1887 and numbered 11301/2. Their original purpose was for transporting standard-gauge stock over the broad gauge. In 1892 the set was converted to standard gauge, renumbered 48999/49000 and coded POLLEN B.

A spate of girder wagon building took place in the first years of the twentieth century: two sets of six-wheel vehicles, similar to the old broad-gauge wagons, were built in 1902 (48979/80 RWA Fig 231 and 48981/2), and a new 40-ton design of four-wheel 20-ton wagon pairs, 39ft 6in over headstocks, coded POLLEN C, was introduced in 1905 (48903/4 RWA Fig 235/7; 48983/4 and 48985/6). During the building period of these wagons the POLLEN A four-wagon set was split up into two-wagon sets. The original 20-ton 'outside' wagons of 12ft wheelbase were paired up, retaining the POLLEN A code, and the 12-ton vehicles of 9ft wheelbase were given buffing ends, uprated to 20 tons and coded POLLEN D. These two vehicles, numbered 32990/1 and built in 1890, were originally described as 'match trucks' (!) for the 20-ton vehicles numbered 32989/92 (RWA Fig 228). The 1905 division of the original four-wagon POLLEN A involved a numbering adjustment, 32989/90 being taken by the 12ft wheelbase wagons and 32991/2 by the 9ft wheelbase wagons. The POLLEN C and the new POLLEN A were similar in appearance with 'lopped off' wheel splashers. In the new design POLLEN C, the newly constructed POLLENS B and the POLLEN A/D rebuilds, the wagons were coupled more closely together than the nineteenth century designs (1ft 1in instead of 2ft), and turntables were installed instead of the original bolsters, thus allowing the wagons to be separated and, for example, shackled to the ends of long bridge girders for easy negotiation of curves (vol.1, p.19).

Left hand end when 4 wagon set as built, 1ft 1in between trucks. Centre wagons simple 14ft oh, 8ft wb.
Right hand end when rebuilt, modified ends etc for 2 wagon set.

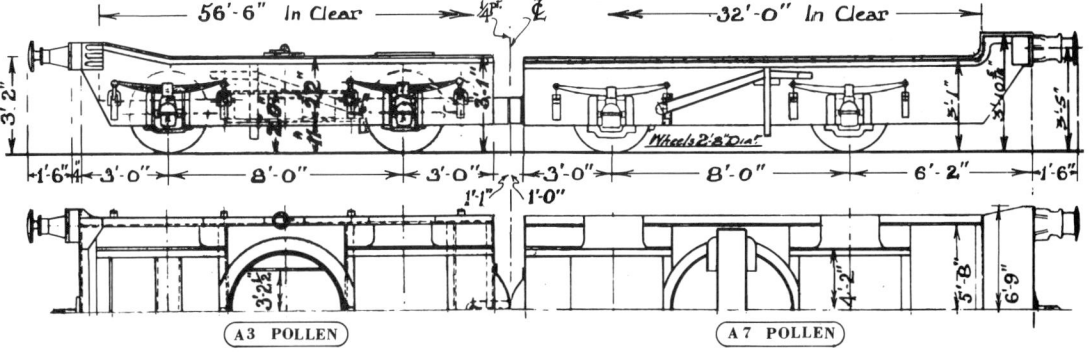

Above: *A4 POLLEN C built 1905. photographed February 1908. Brakes on left hand vehicles in side view. Eyebolt shackles on 3ft springs. Laminated sprung buffing gear. Withdrawn 1936.*

Right: *A7 POLLEN built 1885 as the two outside wagons of the original oldest 4-wagon POLLEN set, converted in 1915–16 into two 2-wagon sets, retaining same telegraph code. Photographed in December 1916; note small GW compared with A4. Old spring shackles on 3ft 6in spring. Patches on sides around wheels. Brakes on right hand vehicles in side view. Self-contained buffers of rebuild. Withdrawn 1936 (cf differences in 36951 CROCODILE C(G) photographed when 4-wagon set, see p. 6).*

The coupling of these wagons was by means of semi-permanent links and central curved buffing bars (cf J9 MITES).

When the wagon diagram index was set up in 1905, in the spirit of 'heaviest, longest, newest' coming first, the following assignments were made: A1 – POLLEN B; A2 – POLLEN A; A3 – POLLEN; A4 – POLLEN C; A5 – POLLEN D. All these vehicles had simple one-sided lever brakes acting on a single-axle, even A4 built at the time Dean-Churchward brakes were coming into use.

Following the building of the Dreadnought battleships in 1904, the GWR introduced at a cost of £1,202 the POLLEN E set for 65-ton gun barrels. Four 6-wheel wagons, numbered 84997-85000, were articulated together to give 82ft 6in over headstocks. The gun barrel was carried in two special cradles adjustably mounted on each end pair of the set (RWA Fig 232/8). Each separate wagon, with DC II brakes, could take 30 tons if the load was spread evenly over all wagons, so the 1910 *Special Wagon Book* illustrates the set without gun cradles for boilers up to 120 tons. The set was given the index number A6, and until the CROCODILE L was built in 1930, it was the longest and heaviest vehicle on the GWR, with tares of 51 tons 5cwt as a gun set and 41 ton 11cwt without the gun cradle.

In 1915, the four-wagon A3 (which had the oldest girder wagons) was rebuilt into two 2-wagon sets with new ends, the capacity of each wagon remaining at 12 tons; the code was also retained as POLLEN for both pairs. No renumbering took place, the 'outer' wagons 32993/6 being paired, and likewise the 'inner' 32994/5; the refurbished 2-wagon sets were given the new diagram number

A7 (RWA Fig 229). The A3 diagram thus vacated was taken by a curious pair of service girder trolleys 33984/5, which had previously been without an index number. These wagons, built under loco department lots, had only a 5ft wheelbase and were only 9ft over headstocks; they had no buffing gear and were rarely used except in Swindon.

Although the gun set A6 could be used for other purposes, no separate diagram was issued for it in the 120-ton guise until the early 1920s when A8 was issued; shortly after, the gun combination was uprated to 100 tons (the photograph in the 1936 edition of 'Exceptional Loads', reproduced in RWA Fig 238, is of a 108-ton gun barrel). To increase the versatility of the A6/8 set, detachable buffing and drawgear was made in 1930 for the centre pair of wagons (84998/9), to allow them to be used as a two-wagon pair carrying 30 tons each; this was diagram A9. The two outer wagons as a pair became A10. Two diagrams were necessary because the geometry of the detachable headstocks on A9 reduced the length in clear from 39ft 6in (A10) to 36ft 6in, although the length over headstocks, 43ft 6in, was the same for both A9 and A10. Although the small service trolleys (A3) were condemned in November 1929, thus vacating A3 again, new diagram numbers were issued for these alternate twin-wagon POLLEN E arrangements. Thus diagrams A6/8-10 all refer to the same vehicles.

In June 1936 all A7 were scrapped and about the same time 48903/4 and 48983/4 of A4 were condemned. In 1938 proposals were put forward to convert some of the articulated wagons into single girder trolleys, but nothing came of it and the rest survived into nationalisation.

CHAPTER 2
B – ARMOUR PLATE, INGOT AND MILL ROLL WAGONS
(including Totems)

Heavy goods of the above types, and similar items taken by POLLENS and CROCODILES could not be taken over long distances by road at the turn of the century. Although the 1900 GWR *Telegraph Message Code Book* alludes to the BEAVER A as intended for roll traffic, when the index was set up in 1905 only two wagons were assigned to the B group, the BEAVER going to the H group. B1 (TOTEM) was a 34ft over headstocks three-bogie 12-wheel armour wagon (41899) rated at 45 tons (Vol.1 p.64). According to the Swindon diagram book it had been built in 1895 from parts of the old 12-wheel broad-gauge engine truck No 11111 dating from the 1870s, although the vehicle is shown in the 1893 *Special Wagon Book*. B2 (TOTEM A) was a shorter 22ft 1in over head-stocks 45 ton bogie roll wagon (41910) built in 1899. Of interest in these untrussed wagons was the deep web I-beams from which the solebars were made, 1ft 1in in the case of B1 and 1ft 4in for B2 (Vol. 1, p.56). The three-bogie TOTEM was later altered to a more conventional two-bogie wagon with adjustable kingpost trussing. A 1922 abridged version of the telegraph code for goods agents still referred to this wagon as being 12-wheeled; however the drawing number of the two-bogie version is 31301 which dates it approximately at 1907 and RWA Fig 33 is dated 1910, so the 1922 list seems in error. The TOTEM B was uprated to 50-ton capacity in the late 1920s.

Just before World War I some old 15ft two-plank open wagons were converted to 10-ton roll wagons, adjustable chocks and bolsters being provided (RWA Fig 55). These rebuilds were given the index B3, and at roughly the same time the G5 SERPENTS A and B (22000 and 21999) were given the same treatment, becoming B5 (20-ton) and B4 (14-ton) respectively. Unusually, these wagons retained their telegraph code names, the descriptions of which had to be altered.

Various Rhymney Railway ingot and roll wagons were taken into GWR stock after the grouping. They did not all immediately receive diagram numbers, but one became B6. It was a six-wheel 40-ton special bolster armour wagon (RR452) that had been converted from an 1872 Sharp, Stewart double-framed 0-6-0ST engine. The GWR coded it TOTEM B and gave it the number 41976 (cf CROCODILE series). In January 1929 the wagon was converted to a 30-ton roll wagon to replace the B4, SERPENT B, which was condemned. Order R22 fitted it with 'permanent packing for conveying heavy rolls'. It lost the TOTEM B coding, though it took over the running number 21999 of the condemned wagon. A group of ex-Rhymney Railway 15-ton 11ft 8in wheelbase wagons was then given the vacated B4 diagram. SERPENT A, B5, followed its counterpart to the scrapyard in 1934, and its replacement was the ex-Rhymney Railway MOREL B, 36976. This wagon was formerly a Rhymney six-wheel well glass wagon (cf numbering of D1), converted in 1919 from the tender of an 1859 0-6-0 goods engine, and used by the GWR as a propeller wagon. It lost its centre pair of wheels in 1934 upon conversion to a roll wagon, although the box well was left intact. As explained by Mountford, the frames of this vehicle were 101 years old when withdrawn in 1966! No diagram had been issued for the MOREL B, but the roll wagon version became B7.

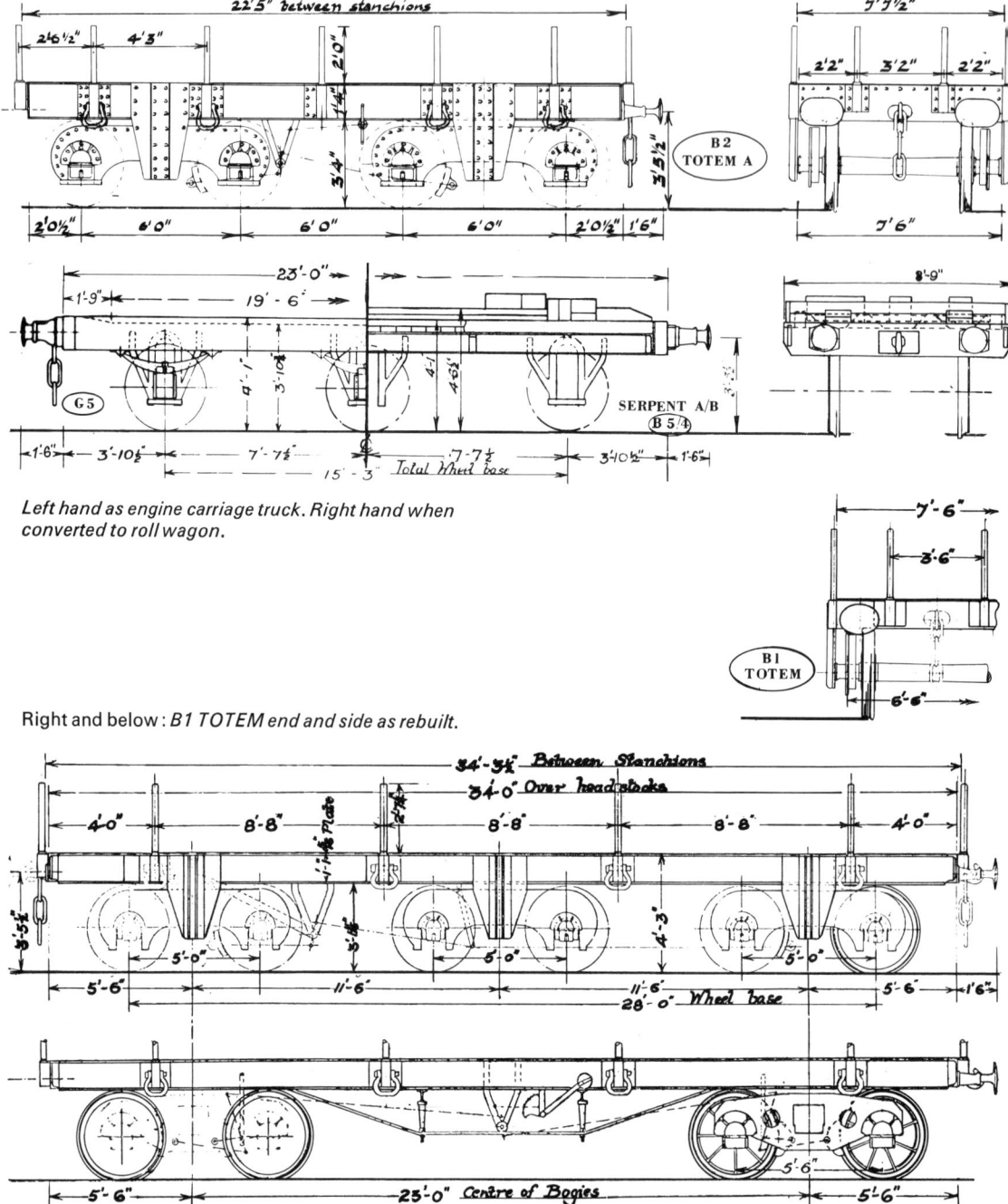

Left hand as engine carriage truck. Right hand when converted to roll wagon.

Right and below: *B1 TOTEM end and side as rebuilt.*

The vacated B5 was taken by ex-Cambrian Railways roll wagons 32183/4 originally assigned 41955/6 at the amalgamation.

Further roll wagons were conversions of OPENS (such as O11 to B8 in 1935 and O32 to B9 in 1939) and the two H3 BEAVERS E to B10 in 1942. B10 retained DC III brakes from the 1911 wagons upon rebuilding even though the 1939 edict about either-side brakes was in force. The thirty-three wagons of B9 were taken out of stock in 1940, and were converted to E4 AERO airscrew vehicles. After World War II B9 reappeared as roll wagons. Roll wagons usually carried the legend 'To be used for heavy Rolls only. Blocks must not be removed from wagon'.

CHAPTER 3
C – BOILER TRUCKS, LATER CALLED TROLLEYS
(Crocodiles)
AND F – STEAM ROLLER TRUCKS

The history of GWR CROCODILES is rather complex, and like the telegraph codes, can be misinterpreted unless the whole story is seen in perspective.

The precursors of modern bogie well wagons were different in appearance from the accepted GWR CROCODILE. The earliest heavy boiler trucks were comparatively stumpy with small enclosed bogies that used large 3ft 7in separately sprung wheels under high end platforms. Two 30-ton boiler trucks had been built in the regular lot series in 1873-5 on osL101, numbered 20449/50. They were the first bogie vehicles ever on the GWR, and although their design is not known with accuracy, their overall dimensions (39ft over headstocks, 20ft well, 7ft 6in wide, 3ft 1¼in high platforms) were somewhat similar to four 35-ton wagons built during 1886-90, 36949-51 and 11169, the last being a 'convertible' broad-gauge wagon. These were followed in 1890 by a similar 40-ton wagon, 41900, which was 10ft shorter in the well (14ft 4in rather than 24ft 4in) but had identical 9ft 4in end platforms. The general outline of these wagons with a 4ft wheelbase bogie is referred to here as design I.

Already in existence at that time were the loco department 4-wheel 34ft over headstocks 15-ton boiler trucks, which had been built separately from the merchandise stock by the loco department itself. One similar wagon was 47ft over headstocks with a 40ft wheelbase, the others having 28ft wheelbase. These four-wheelers were easily the longest rigid wheelbase vehicles ever on the GWR,

the maximum fixed wheelbase of four- or six-wheeled carriages being 22ft.

There is evidence that when it was outshopped in 1890, the third standard-gauge 35-ton trolley 36951 had an extra 6in on each end platform to accommodate 4ft 6in wheelbase bogies; 36949/50 were soon similarly altered, but the former broad-gauge wagon, renumbered 41901, retained 4ft bogies for some years in the 1890s before being given the lengthened 4ft 6in version. The outline of lengthened wagons with large wheel 4ft 6in bogies is referred to here as design II.

In 1891 what amounted to a bogie version of the 47ft four-wheeler appeared, rated at 20 tons (design III), vol.1, p.59. The well was comparable in length but the end platforms were 8ft 6in long as opposed to 6ft 9in, giving 50ft over headstocks; 41902 was quite rakish in appearance, the skirts below the end platforms being flared off nicely. The new 4ft 6in bogie with 3ft 1in wheels was the first plateframe design on the GWR and, like centreless scroll-iron carriage bogies, was suspended from the side skirts and frames, not from the pivot boss. Holes in the skirts gave the rear axleboxes clearance on sharp curves and although

C3 CROCODILE B (old code). Built 1891, photographed 1898. Design III. 16 by 6in RSJ well girders, end plates. First plateframe bogies on GWR, 4ft 6in wb, suspended from wagon side skirts. 'Round' 8 by 4in axleboxes. Single lever brake on one bogie. Oval buffers, round guides. White patch on label box for card frame. Plate on left hand end same as B2 TOTEM A, Vol 1, p. 56. No instructions on ends of platforms or well girders regarding loading.

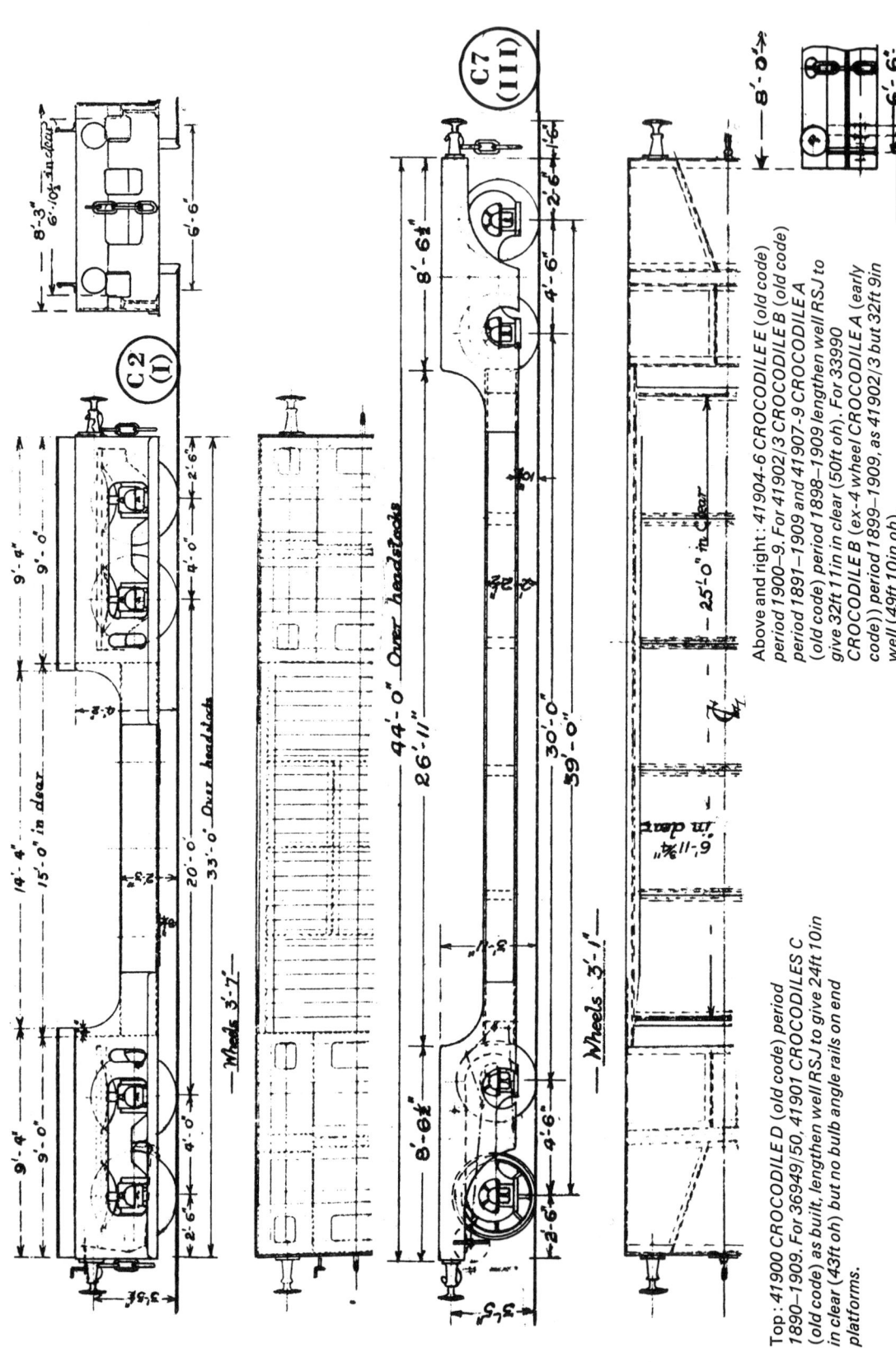

Top: 41900 CROCODILE D (old code) period 1890–1909. For 36949/50, 41901 CROCODILES C (old code) as built, lengthen well RSJ to give 24ft 10in in clear (43ft oh) but no bulb angle rails on end platforms.

Above and right: 41904-6 CROCODILE E (old code) period 1900–9. For 41902/3 CROCODILE B (old code) period 1891–1909 and 41907-9 CROCODILE A (old code) period 1898–1909 lengthen well RSJ to give 32ft 11in in clear (50ft oh). For 33990 CROCODILE B (ex-4 wheel CROCODILE A (early code)) period 1899–1909, as 41902/3 but 32ft 9in well (49ft 10in oh).

clearly off-centre when the vehicle was straight, afforded plenty of access for oiling. When the telegraph code for wagons was introduced in 1892, the four-wheelers became CROCODILE (34ft) and CROCODILE A (47ft), and the bogie wagons CROCODILE B to D in increasing load capacity. The 20-ton CROCODILE B was joined in 1896 by 41903 to the same design, and in 1899 the four-wheel CROCODILE A was rebuilt to a 20-ton bogie CROCODILE B by the addition of the end platforms; the well was slightly shorter giving 49ft 10in over headstocks. At this time new bogie CROCODILE categories appeared. Both the CROCODILE E (41904-6) and the (new) CROCODILE A (41907-9) were similar in design to the CROCODILE B, but were rated at 25 tons, 8in by 4½in journals replacing 8in by 4in; the CROCODILE A was 50ft over headstocks like the CROCODILE B, but the CROCODILE E had a shorter well to give 44ft over headstocks. Also at this time new four-wheel CROCODILE wagons were added: 33988/9 (loco department) and 41911-14 (ordinary stock). The brakes on the earliest CROCODILES were screw type with the crank on the buffer beam (vol.1, p.58 and rear dust jacket of RW), and even between the buffers on some bogie boiler trucks. This arrangement developed into the Thomas brake proper on the side of the vehicle.

In 1897 a 25-ton bogie well wagon, 41000, was built for the construction department to carry traction engines or steam rollers and their water carts. The vehicle resembled the contemporary design III CROCODILE E, with ramps leading down to the well. Three further vehicles (40996/8/9) followed in 1900-1, and all later became F2 in the index. The 'missing' 40997 can be traced to L274 which was cancelled; however, the following wagon, built on L353, took its originally allocated number, 40997 being allocated to the first GANE (later J1) built on L258.

L378 of 1902-4 introduced the mutation in designs between old and new bogie CROCODILES. Four vehicles were built to carry 25 tons (40 over bogies). They had the same well of the CROCODILE A wagons of a year or so earlier, but used 5ft 6in plateframe bogies, which required an extra 1ft on the end platforms to give 52ft over headstocks. For the first time the side plates, at the junction of well and end platform, were sloped. All bogies were suspended centrally (not from the frames) so the skirts under the platforms could be removed, as in the CROCODILE G, 41916-8, the profile of which is referred to here as design V (RWA Fig 106). The first

wagon of the lot, CROCODILE F 41915, retained the skirts, using a short-lived bogie design (the wagon outline being called here design IV); a contemporary 4ft 6in version of the bogie may be found on F1, but the others used the 5ft 6in design then being applied to MACAWS B, MINKS F and bogie coal wagons. DC brakes appear for the first time on designs IV and V acting on one bogie.

The years 1903-4 saw two new road roller wagons (F1); like 41915-18, design V, the bogies were centrally sprung, so the skirts disappeared. The well at 3ft 6in was higher from the railhead than F2, some of which were later rebuilt to conform. The ramp slope may have been too steep for comfortable loading.

Two service four-wheel boiler trucks, 33986/7, had been constructed for the locomotive factory, so when the diagram index was set up there were ten distinct boiler wagon designs; their sequencing does not obey the 'newest, longest, heaviest and widest' principle too well.

Extensive reconstruction to design V of all the old bogie vehicles built before 1902 took place in the period 1902-8, some being downrated in the process; 36949-51 were rebuilt with 5ft 6in bogies, becoming 47ft over headstocks. The first new wagons added to the C index (C11 CROCODILES B) in 1905 also followed design V. However a radical change in construction methods appeared on the C12 CROCODILE G in 1908. Instead of using structural girders for the well, with cut-out plates for the join between well and end platforms, the whole side profile was cut from plate and made into built-up I-beams by the addition of riveted flanges. Angle sections thus appear all along the wagons in both elevation and plan, not only in the well. The typical profile is called here design VI and lasted into nationalisation. Similar changes occurred with some other well wagons, as explained in Chapter 4. A further distinguishing feature between designs V and VI was the width of the 'neck' between well and end platform. The inside wheels of the bogies were partially covered by a wide neck in the case of V, but on most VI designs the neck was much narrower. C12 also lightened the wagon with holes along the solebars, but apart from the rebuild of 33990 (C6) in 1932 the practice was not followed later. Although new construction augmented existing wagons as regards vehicle specification and capacity, new diagram numbers were issued to distinguish design VI vehicles from V (C12-14); vol.1, p.59, RWA Fig 35. C15 was a new departure, this 20-ton wagon having a 40ft 7½in well (62ft over headstocks), vol.1, p.20; identical end platforms to C13/14 were used (RWA Fig 32).

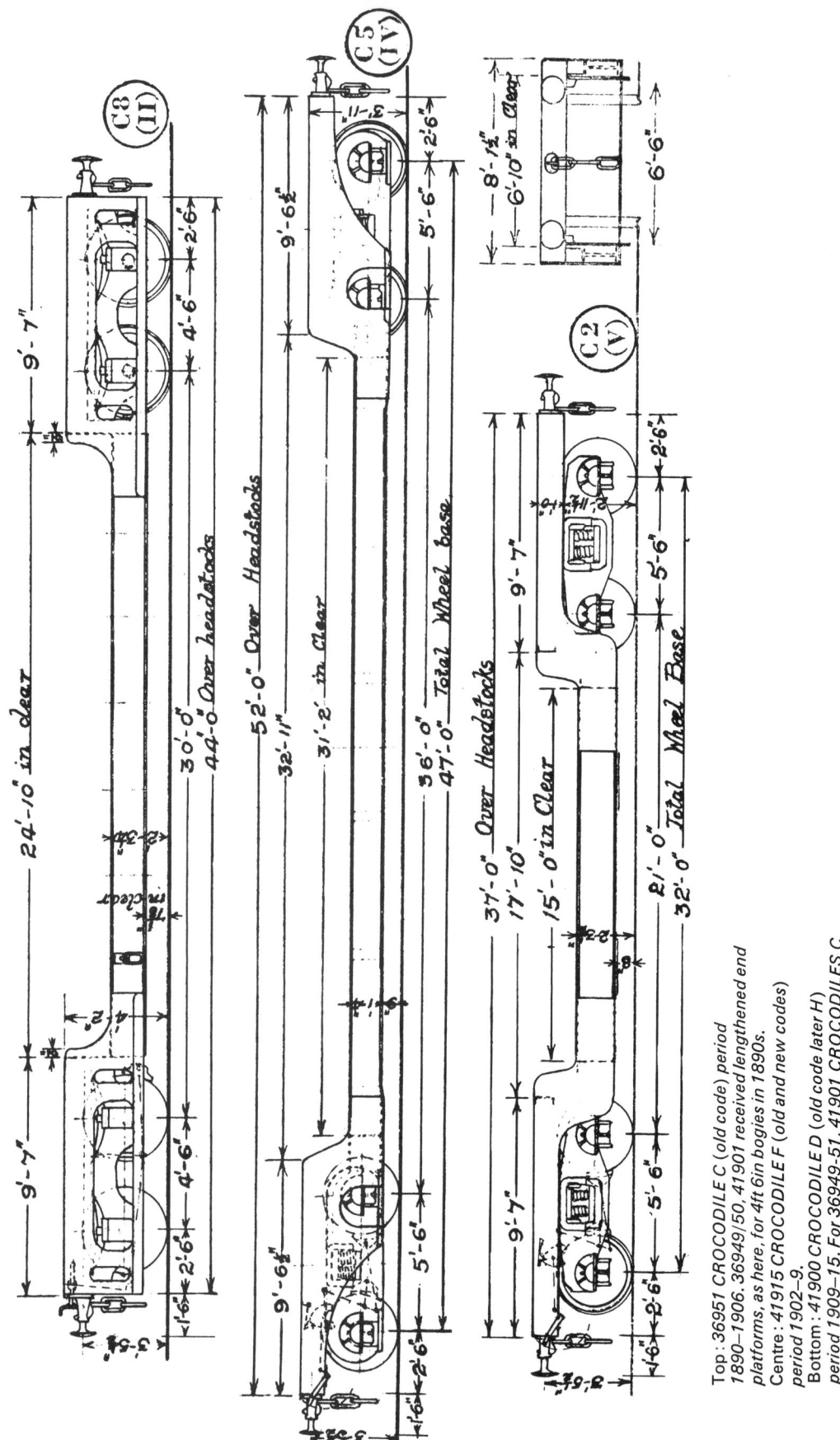

Top: 36951 CROCODILE C (old code) period
1890–1906. 36949/50, 41901 received lengthened end
platforms, as here, for 4ft 6in bogies in 1890s.
Centre: 41915 CROCODILE F (old and new codes)
period 1902–9.
Bottom: 41900 CROCODILE D (old code later H)
period 1909–15. For 36949-51, 41901 CROCODILES C
as altered after 1906, add end platforms as here to
24ft 10in well RSJs to give 47ft 0h. Other types V
similar.

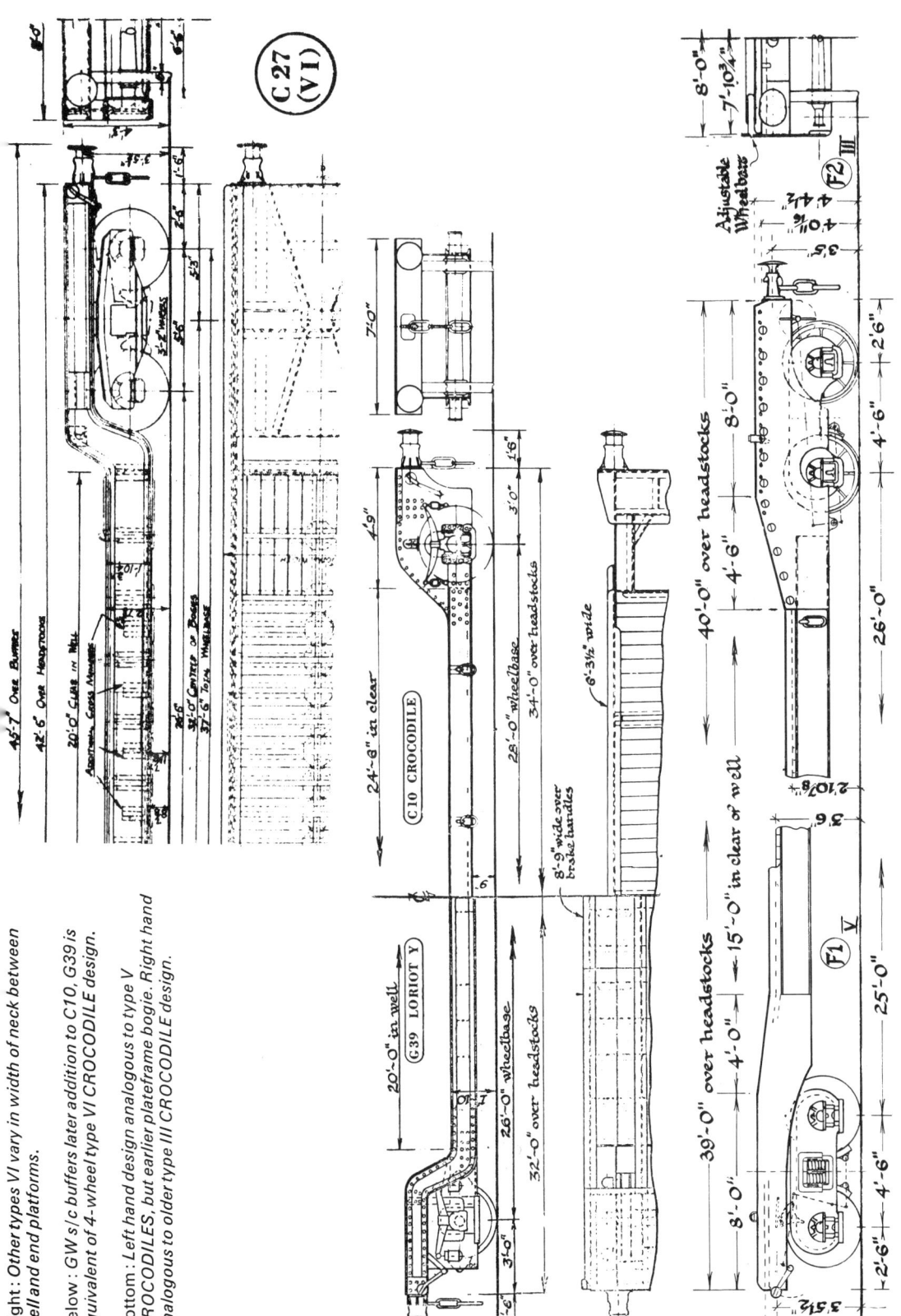

Right: Other types VI vary in width of neck between well and end platforms.

Below: GW s/c buffers later addition to C10, G39 is equivalent of 4-wheel type VI CROCODILE design.

Bottom: Left hand design analogous to type V CROCODILES, but earlier plateframe bogie. Right hand analogous to older type III CROCODILE design.

C 27 (VI)

C10 CROCODILE

G39 LORIOT Y

F2 III

F1 V

C10. CROCODILE built and photographed 1898
(outside Newburn House, home of CME). D-shackles,
10 by 6in OK oilboxes, ten-spoked wheels. Thomas
brake. Unusual swivel coupling hook. Plates as on B2.
Later modifications included self-contained buffers
and drawgear, DC II brake, eyebolt shackles. Note
construction method of joining plateframe ends to
I-beams in well. Note join of chaired rail to baulk track.

The telegraph codes for CROCODILES and LORIOTS were changed in summer 1909 to bring logical wagon size variations in as well as capacity. Table 7 in Chapter 3 in Vol. 1 shows the reallocations, as well as certain transfers to different diagram numbers which also took place at this time with rebuilds. One such alteration concerned 36949-51 in C8; these had been rebuilt from design II to V in 1905, but were strengthened with 4in patch angles on the sides of the end platforms in the years following 1908, and one by one were transferred to the new diagram C16. They could be mistaken for design VI, except that there was no built-up flange in the neck between well and platform (RW p.70). Again, in plan view the patch angles projected out so that C16 was 8ft 9½in as opposed to 8ft 1½in at the platforms of C1. Some other old wagons rebuilt to design V were subsequently given design VI built-up end platforms in the 1930s.

C17 was an interesting light trolley built in 1909 for the conveyance of motor vehicles. The floor of the 25ft 6in well could be raised level with the end platforms, so that the vehicle could be driven on. Lowering the floor down into the well allowed motor buses to clear the loading gauge, as the floor was only 11½in above rail level. Contemporary LORIOTS and HYDRAS had floors about 2ft 6in above rail level, and even though the 'curvy' G17 LORIOT was lower at 1ft 3in, the end ramps were quite steep; moreover, no four-wheel vehicle had such a long well. A drawing at Swindon of C17 in its 'proposed' form shows the vehicle as 8 ton, 48ft over headstocks, vacuum-fitted and using 5ft 6in bogies. It was built however at 10 ton and 46ft over headstocks, using the 4ft 6in plateframe bogie as on C11 (RW p.72). The wagon was called CROCODILE K after the code reorganisation.

Five more service boiler trucks appeared in 1912-13; their dimensions differed slightly from earlier vehicles and the new diagram C18 was issued. Six smaller boiler trucks for which no diagram was issued had been built in 1909 (L622) from condemned tenders. They were originally numbered 33945-50, but immediately were altered to 800-6; 33946-50 were then taken by the new C18 vehicles. Eventually the earlier C9 vehicles were transferred to the C18 diagram in the early 1920s.

Two further wagons were built in 1913 to carry scarifying steam rollers. Their design bears to F1 what the bogie CROCODILE design VI bears to V. Angles appear all the way along, and the modern 5ft 6in bogie was used. Both wagons were originally given F3, but F4 was issued for 40992 when it was renumbered 42031 (formerly belonging to a condemned G2 LORIOT B) and transferred into merchandise stock in the late 1920s with the new code LORIOT R.

World War I saw various CROCODILES uprated and others prepared for military use. Three CROCODILES G were built under Order O.387 in August 1915 to carry 6in Admiralty guns and had the war diagram C19; logically they were 41952-4, but were not in GWR stock. Screw couplings and safety links for abroad were fitted to 41906 in September 1915 and was sold to the government. Up to this time large capacity CROCODILES (40-45 ton) were either 42ft 6in or 47ft over headstocks. The 40-ton CROCODILE J of 1916 was a longer vehicle, 54ft over headstocks. The end platforms were slightly longer than normal at 9ft 9in, but the 31ft 6in length in clear at the well was greater than any wagon of comparable capacity. The platforms

C11. CROCODILE B (old and new codes) built 1906, photographed 1934 en route Chepstow/Newport Docks. Design V, with 4ft 6in wb single visible spring bogies, round 8 by 4in axleboxes. Inside wheels partially covered by plates joining end platforms to well. Square shank oval buffers, ₵ mark, loading instructions, Not Common User. Label clips as well as box.

were also higher than usual at 4ft 5½in necessitating a deep buffer beam and unusual tapered side profile (RWA Fig 34); along with J19, it was the first bogie vehicle to use large circular buffers instead of oval. This wagon re-used the number 41906 of the sold CROCODILE A, and took the C8 diagram vacant from the conversions of 36949-51 to C16. This was probably the first instance of re-use of a vacant diagram (cf G diagrams later). Some four-wheel CROCODILES were equipped with GWR self-contained buffing and drawgear during and after World War I; since the transverse laminated spring was absent, a stiffening bar was attached from the drawgear down to the well (compare the two photographs in RW p. 64).

Another war diagram, C20, was issued for the RECTANKS built by the GWR. New Work Order 1016 of April 1918 concerned the building of 40 35-ton trolleys for tanks to Midland Railway drawings, the original R(ailway) E(xecutive) C(ommittee) Tank Carrier design with jacks (BGW p. 112); their numbers were WD 12041-80. Along with some overseas RECTANKS they returned to Swindon after the war and were modified for GWR use in 1920 (the brakeshaft was altered so that a 1½ chain curve could be negotiated, and the corner jacks were removed); diagram C21 was given out (Vol.1, p.25). They were still technically on loan from the War Department and were given the GWR hire nos T06369-408; WD7/8/10 from elsewhere became T06673-5. Then, in August 1921, juxtaposed with the order for the J23 MACAWS E, nine RECTANKS were overhauled and given GWR numbers 70745-53; similarly, L892 of February 1922 marked the purchase of the other 31 RECTANKS built by the GWR for the government, for which the running numbers

17310-40 were assigned. Next year, bolsters were fitted to some of them (in 1919 similar fitments had been applied at Swindon to other RECTANKS for the North Eastern Railway). These three-bolster wagons (for which no diagram was issued) were later allocated to Port Talbot docks, and should not be confused with the four-bolster wagons converted after World War II (J31). Along with the ex-TVR bolster wagons coded MACAW G (no diagram), the RECTANKS were the only GWR vehicles to have diamond-frame bogies. RECTANKS 17310/13/19/20/4/48 were strengthened later to carry 20-ton steam rollers and traction engines.

No C19 vehicles were returned by the government, so C22 was issued in 1921 for replacement vehicles which took the original numbers 41952-4; likewise a new CROCODILE A was built in 1923 to replace the old 41906. The new vehicle (of design VI instead of the design V it replaced) took the 'new' C13 diagram which had been vacated upon the uprating and transfer of 41947 to C2. Since the old running number had already been taken by the CROCODILE J, the wagon was given the next available number 41956, 41955 having been taken by the second CROCODILE J built in 1923-5.* Further re-use of vacant diagrams occurred when some new 62ft CROCODILE E wagons built in 1923-5 were given C9, all the service boiler trucks having been grouped into C18. Again C19 and C20 were given to new CROCODILES G and F respectively. Thus, although CROCODILES E, F and G of design VI were

*41955-6 were the numbers originally assigned to ex-Cambrian Railways well wagons nos 2510-11; they were however almost immediately converted into roll wagons (print 65923) under Order 4089 in 1923. Compare also 41976, TOTEM B.

C13 CROCODILE D (old code and index; later C2 CROCODILE H). Built 1908 photographed March 1909. Design VI, 37ft oh. Loading instructions on end platforms. Only bar on brake handle painted white. No external stiffeners on bogies; two visible springs.

C2 CROCODILE H photographed 1940. Same vehicle as old C13 strengthened to 65 tons in 1936 (42ft 6in length of 1915 rebuild to 45 tons), 5ft 6in plateframe bogie with external stiffeners; single spring visible. Note different positioning of lettering from other picture.

already in existence, new diagrams were assigned. The explanation of this is that Swindon tended to issue new diagrams for quite small constructional changes. Here, the differences related to self-contained buffing and drawgear; C23 of 1926 was a similarly updated version of C2.

The early 1930s saw more reconstruction, the fifty year old 33990 being given new design VI end platforms attached to the retained original I-beam well; the old drawgear was also retained so that the square-shank Churchward buffers remained. Holes were drilled in the beams to lighten the tare, and the load of the wagon was increased to 25 tons. Since contemporary built-up riveted end platforms, as on C23, were longer than the old plate end platforms, this 1930s rebuild gave a longer wagon, 53ft over headstocks (RW p.67). The new number 41978 was assigned to C6, and those of C10 left in the 33xxx/34000 series were renumbered to 41979-86 at the same time. The CROCODILES C/F C3, that had been built originally to design III in 1891-6 and subsequently altered to design V about 1907, were likewise reconstructed with design VI end platforms in 1935 (53ft over head-stocks), as were 41915-16 (54ft over headstocks). The latter were also given electrically welded floor plates, which gave an extra 3¾in in height from floor to top of load gauge, and two slide bars across the well. Again, in 1936 under Order 3163, all the

F2 built and photographed 1897. Thomas brake on one bogie, early 4ft 6in plateframe design suspended from side skirts. Note connection between well girder and ends. Round guide oval buffers. Design is analogous to type III CROCODILES. Italic script says 'Engineering Department for Carrying Steam Roller and Water Cart'.

F3/4 (later LORIOT R) built and photographed 1913. 5ft 6in bogie with external stiffeners and 'square' 8 by 4½in oilboxes. Label box has two card frames. Brake handle painted, square shank oval buffers. Design is analogous to type VI CROCODILES. Circa 1928, this wagon was renumbered 42031 (ex LORIOT B, G2), coded LORIOT R, and put into merchandise stock as F4.

CROCODILES H were strengthened to take 65 tons, 'single' coil bogies being used (RWA Fig 36). Likewise about the same time the old broad-gauge CROCODILE D (C1) which had been downrated in 1909, unlike its contemporaries 36949-51, was brought up to 35 tons and recoded G. The D code thus was vacated (but was not taken up by C28, see later).

In 1926 the Electricity (Supply) Act introduced the National Electricity Scheme, forerunner of the present-day grid system, and there arose a need for vehicles to carry transformers. The G23 10-ton MAYFLY wagons had been introduced in 1919 but the size and weight of transformers grew tremendously in the following years, and L1025 of 1928 was issued for an 80-ton trolley to which the running number 41977 was given. Well girders were to be slung between two pairs of 5ft 6in plate bogies at each end, giving a 16-wheel wagon. This lot was cancelled, however, and in 1929-30 a massive 120-ton trolley with detachable inter-changeable side girders (straight and well types vol. 1, p.61) was constructed instead under L1042, keeping the original running number. The vehicle, coded CROCODILE L and given C24/5, was like an expanded 24-wheel POLLEN in appearance, (see BGW p.113 and RWA Figs 112/4). If rails were laid into a newly built power house, the wagon was run in and the transformer or stator jacked up. The side girders of the wagon, now released of their load, were withdrawn and the machinery lowered into place. Additional cross-members were fitted at this time to 41961/4 of C19 and to 41973/4 of C23, also to facilitate loading transformers. For these modified wagons C26/7 respectively were issued in 1931, although Order R3 of 1928 actually introduced the idea. Fig 36 in RWA is of C27 not C23.

Just before World War II, four long trolleys coded CROCODILE M (C28) were built for structural steelwork; 62ft 6in long, they could carry 12 tons (20 over bogies); RWA Fig 110. Although they were 1ft longer in the well than the 62ft CROCODILE E, because of 9ft 4in rather than 9ft 7in end platforms, it is not clear why the CROCODILE E design could not have been used with smaller bogie journals.

In connection with the 'state of readiness' before World War II, extra lashing rings for two tanks were provided on RECTANKS vol.1, p.25, and the Port Talbot bolster wagons were reconverted. Then in June 1940, twelve extra RECTANKS were obtained from the Royal Ordnance Depot at Didcot in exchange for 16 OPENS C on a mutual hire basis. The WD numbers were prefixed by the figure 2, giving 212005/81/2/9/94/5/107/8/12/5/8/20 (RWA Fig 101). Early in the war 36949, 41934/51 and RECTANKS 17318/30/5/7/9/46/50/3 were lost in France, and 41953-4 were sold to the Admiralty for AA gun use in September 1941 (cf original 41952-4 of World War I, Vol.1, p.25). Nos 41953/4 were however soon returned to GWR service in March 1942. Nos 17330/5/7 and 41934 were recovered and returned to service in August 1945. A Swindon photograph shows the CROCODILE F at Calais after capture from the Germans, who rated it, according to the stencilled figures, at only 20,000kg (20 tons). This wagon is now preserved at Didcot.

After the war, two C9 CROCODILES E, 41957/8, were strengthened to take 30 tons distributed in the well, rather than the previous 20 ton limitation. This gave rise to the C29 diagram in 1945 (cf legends on wagons in Figs 115/7 in RWA).

As for the floors and plan views of the CROCODILES, the end platforms were always of steel plate sometimes with triangular openings to the track, and most wagons had conventional cross-floorboards in the wells; later wagons had longitudinal floor planking. A combination of longitudinal planking, and cross planking on the end ramps, chocks was used on the CROCODILE K and RECTANKS. Construction just before the turn of the century used steel plate in the well; examples were 33990 and 41902-9 in their original design II guise, but they received cross floorboards on rebuilding. An iron floor, level with the underside of the solebar, was re-installed in 41904 in the summer of 1928 (Order R8), thus making it suitable for conveyance of large ships' propellers. The official 'Exceptional Loads' photograph (RW p.66) of a ship's propeller en route from Brentford to Barry, was taken at Cardiff in 1920; 41904 there is thus in unmodified condition. Plate floors were provided on 41915/6/52-4/7-60, so that only the two original C15 CROCODILES E had planked well decks. Deep cross-members were fitted during the rebuilding of 36949 before World War I.

CHAPTER 4
D – WAGONS FOR PLATE GLASS
(Corals)

The essential features of these wagons were racks, in which large plates of glass could be stacked vertically, and a well to accommodate them. To allow an unobstructed well, the solebars were made of deep stiff sheets of iron, the base of the well then being determined merely by axle height (giving about 2ft above the railhead).

The first such wagon was built in 1884 and had a well and racks 16ft 6in in clear. This was 36970, but the remaining vehicles built under the same lot, 36971-5, had a 17ft 6in well, making them 20ft over headstocks instead of 18ft. The 19ft wheelbase and 12-ton load were common to all; a simple one-sided lever brake with shoes acting on one wheel

was used. The diagram D1 (with the 17ft 6in dimension) was eventually issued for these wagons, which were coded CORAL. More glass wagons were built in 1899 and 1908. These too had 17ft 6in by 4ft in clear well, but the two inner of the four partition racks were adjustable (CORAL A). The frames of this design (D2) were cut out in smooth curves rather than the angular appearance of D1, and the deck of the wagons had an inflexion over the axleboxes not seen in D1, to clear the 3ft 1½in wheels; D1 used 2ft 8in wheels. The 1899 lot were equipped with the Thomas brake (vol.1, p.56), but eventually Dean-Churchward replacements appeared. One other difference between the

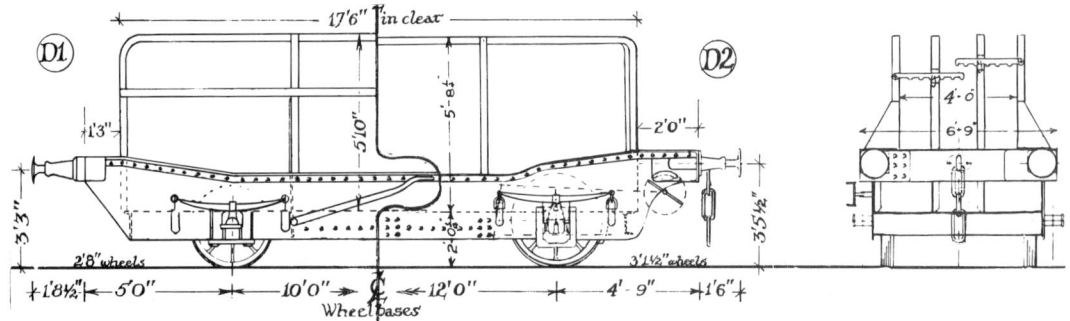

*D2 CORAL A built 1908, photographed 1941.
Eyebolt shackles, disc wheels. DC brake (central
diagonal stripe). Note differences from other
photograph, vol 1, p.56, riveting, keepers below
axleboxes, spring stop design. As a plateframe vehicle,
16in lettering is retained post 1936.*

two lots of D2 was the suspension, the earlier wagons having D-shackles, the later eyebolts.

CORALS were narrow wagons, and their buffers (set in line with side frames) were 'right on the edge' of the 6ft 9in buffer beams. In plan, the deck of D2 'flared out' to the ends of the buffer beam; the ends of the large transverse buffing spring just protruded through the frames, clearance holes being provided, as in some other narrow vehicles. The widest part of the wagon was over the axleboxes and brakehandle.

The GWR inherited a six-wheel glass wagon from the Rhymney Railway, but as explained in Chapters 2 and 5, this wagon was converted into a propeller wagon and later into a roll wagon, although since it took the sequential number 36976 of the wagons in D1, it may have been intended originally to use it as a glass wagon.

36973 was scrapped in the early 1930s; the remaining vehicles of D1 were allocated to Port Talbot for steel plate traffic, together with some of D2 (such as 41712 in 1935). All the 20ft wagons of D1 were scrapped in January 1947, the 'odd' wagon of osL361 survived nationalisation.

CHAPTER 5
E – WHEEL AND PROPELLER WAGONS
(Morels, Aero)

In the nineteenth century the 'propellers' for which these wagons were intended were ships' screws and the 'wheels' were typically the large flywheels of stationary mill engines, then powering most factories, or pithead wheels.

The MORELS basically were flat wagons with an open well, having a large pair of trunnions from which to support the propeller or wheel by the axle. The centreline height of the trunnions was some 7ft, thus allowing wheels of perhaps 13ft diameter to be carried within the load gauge, or even 25ft diameter half wheels. The well of the wagon was designed with a steel lip on which slid steel-reinforced wooden chocks, similar to the arrangements on some SERPENTS C/D; curb rails 4½in high were also fitted around the wagon itself. Although originally of 20-ton capacity, loads

suspended directly from the trunnions without packing dunnage were not to exceed 5 tons.

Two lots in the old series, 425 and 456, were issued in 1888 for the construction of the earliest of these vehicles. A 20-ton propeller truck, 42000, which later became E1, was built under osL425; osL456 produced two vehicles 'as lot 425'. Whereas 41999 indeed had the same dimensions as 42000 (25ft over headstocks, 16ft wheelbase, 20ft 11½in by 4ft 2½in well), 41998 was a smaller vehicle (23ft over headstocks, 14ft wheelbase, 17ft 6in by 4ft well); all wagons had 3ft 7in 12-spoke wheels.* The smaller wagon eventually became E2. According to the 1889 *Special Wagon Book* all three

*There is some confusion over these lots. osL425 may have built two wagons, and osL456 one 'to corrected figures'.

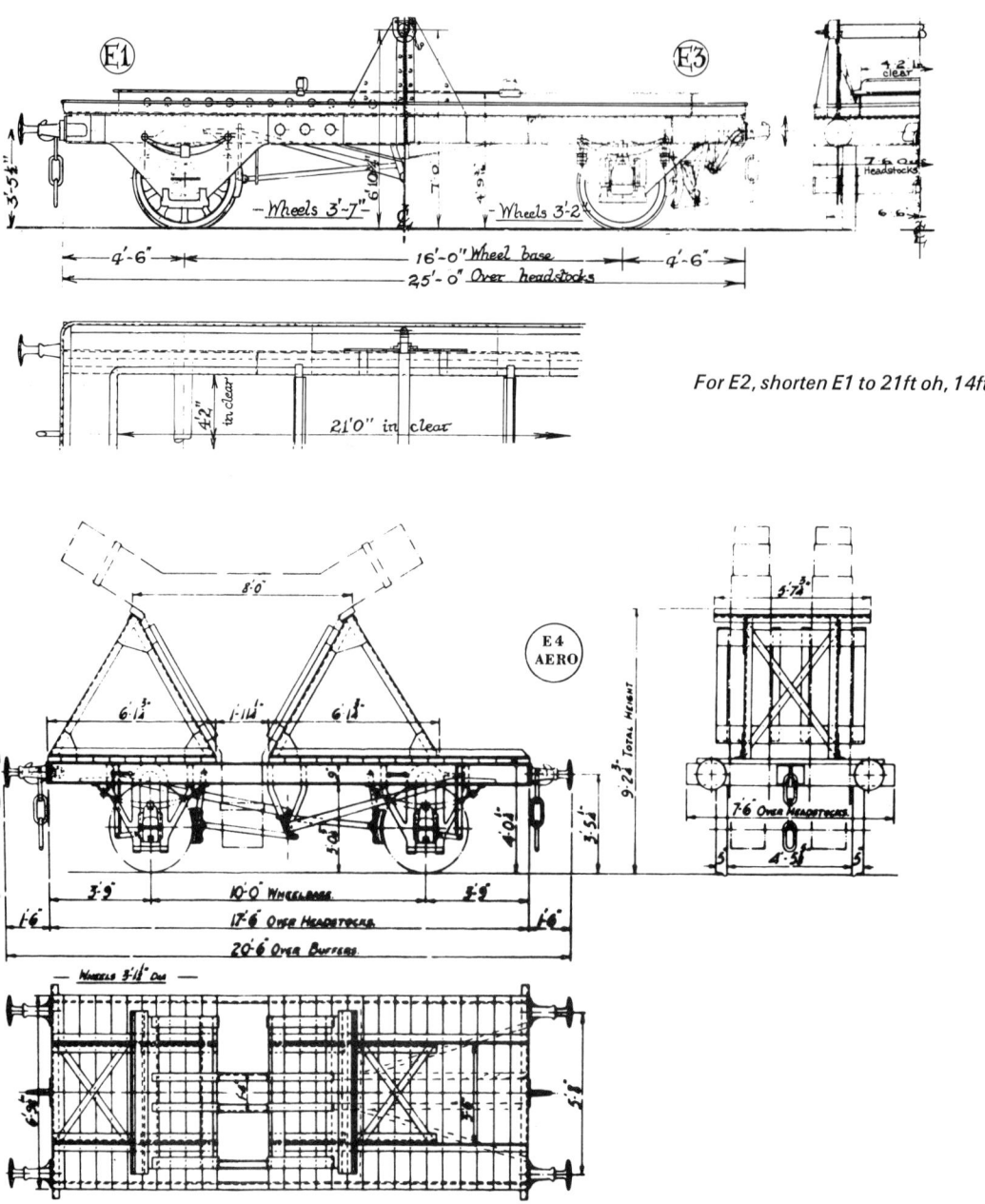

For E2, shorten E1 to 21ft oh, 14ft wb.

were to E2 specifications, but the 1893 book acknowledged that 41999/42000 were different, although only a sketch of the smaller design was shown. Brakes were simple one-sided lever brakes, but in 1909 a DC II arrangement was used on a new MOREL, 41997, which was given the E3 diagram; it was basically similar to E1, but had 3ft 2in wheels and was uprated to 25 tons. The other three wagons were soon also increased in capacity. One of E1, 41999, was converted to DC brakes under

the same L621 that built 41997; it may have been new construction, but that is unlikely.

A 1900 Barry Railway propeller wagon (643) came into GWR stock at the amalgamation. It is not known whether it was reconstructed (and since no diagram number was issued it rather seems not). but it was remarkably like E1 in GWR days. Given the number 41994, it was coded MOREL A (41995/6 had been taken by the BEAVERS E). The Rhymney six-wheel glass wagon 36976 was

E1 MOREL built 1888, photographed 1890s. Eyebolt shackles, 3ft springs. Grease boxes, 12-spoke wheels. Turton self-contained buffers. Central white stripe does not indicate cross-cornered brakes, but does indicate side on which brake lever lies, rather than middle of wagon, as thought originally. Thin 3-link couplings.

E3 MOREL converted 1909 from E1 with DC II brake. Photographed 1922. 3ft 2in 8-spoke heavy wheels. Square shank buffers—spring protrudes through plateframes. 10 by 6in OK boxes.

also initially used as a wheel wagon, and was coded MOREL B. In 1934 however it was rebuilt to a roll wagon, the original box well being retained.

Two-bladed aircraft propellers could easily be carried on ordinary wagons, but in the late 1930s, three-bladed propellers were developed, with adjustable-pitch attachments in the bosses, which required special vehicles. Thus in 1938 five special wagons E4 were introduced for this purpose. Very simply, they were O32 underframes with complete

overall decking except for a small opening; triangular cradles could take two propellers side by side. 170 more were constructed by the end of 1941, the urgency of building being reflected in the fact that L1387 built 50 wagons in 18 days, and L1406 built 100 such vehicles in 45 days! They survived into nationalisation, but in the early 1950s were converted to open goods wagons, O32. Some B9 roll wagons had also been utilised for underframes and were transferred back after the war.

CHAPTER 6
G – FLAT AND WELL WAGONS FOR CARRYING ROAD VEHICLES, COVERED MOTOR CAR TRUCKS AND COVERED TRUCKS FOR MOTOR CAR BODIES

The G index originally included only wagons for road vehicles and agricultural machinery (LORIOTS, HYDRAS and SERPENTS), but in 1919 the MAYFLY transformer wagon was included, and later covered vans for motor cars and motor car bodies were added (ASMOS, DAMOS, MOGOS and BOCARS). New LORIOTS were added to the list after the grouping, including some Welsh wagons, and low index numbers were then re-used in the G group.

Loriots, Hydras and Serpents

Early descriptions of the wagons, which bring out their use, were 'agricultural implement wagons' or 'machine trucks' for LORIOTS, 'road vehicle wagons' or 'tramcar trucks' for HYDRAS, and (horsedrawn) 'furniture van trucks' or 'goods carriage trucks' for SERPENTS; although the vehicles were used interchangeably for some traffic. The first two code groups were essentially well-trucks in modern nomenclature with sloping ramps on the ends: HYDRAS were 'passenger' LORIOTS, with carriage wheels and fittings, which carried smaller loads (usually less than 10 tons) as opposed to the 10- to 20-ton capacity of most LORIOTS. SERPENTS on the other hand were always flat wagons (like later CONFLATS), with no pronounced well; they sometimes had

miniature end ramps but these acted more as chocks. In the nineteenth century some of the earliest LORIOTS and HYDRAS looked 'flat' in profile, but they had ramps hidden behind the side rails. It had become clear by the time of drawing up the index that the code LORIOT was to apply only to wagons with pronounced wells, so that in 1906 the last 'flat-sided' LORIOT, G7 31307 built in 1880, was recoded SERPENT. All types were capable of being end-loaded and all tended to have low buffer heights, even HYDRAS with Mansell wheels. Some later LORIOTS (W and Y) however had no ramps being more akin to CROCODILES (cf G27/39/41) and indeed are called 'trolleys' in the lots. It is significant that their letter sequence was out of order (LORIOTS N and P were conventional well-wagons and were introduced after W and Y), indicating that they were 'odd' LORIOTS (cf also LORIOT R and MACAW Z as being atypical for their codes).

The first true well-wagons were built under 'ragbag' osL214 in 1880, eventually becoming G3 and G4. They had no end platforms, the ramps starting at the headstocks; G4 was essentially a stretched-out version of G3. Subsequent designs between 1890 and 1927 incorporated flat end platforms, 1ft 10in long for G2/1 (26ft 6in over headstocks) and 2ft 1in for G18/20/14 (27ft over headstocks); all the vehicles had 15ft wells and 21ft

For G2, widen G1 to 8ft 6in. For G4, lengthen G3 to 25ft 6in oh, 20ft wb. For G18, narrow G14 to 8ft wide, GW laminated spring buffers, clasp-type shoe on far wheels (as on G14). For G20, bring G18 to width of G14.

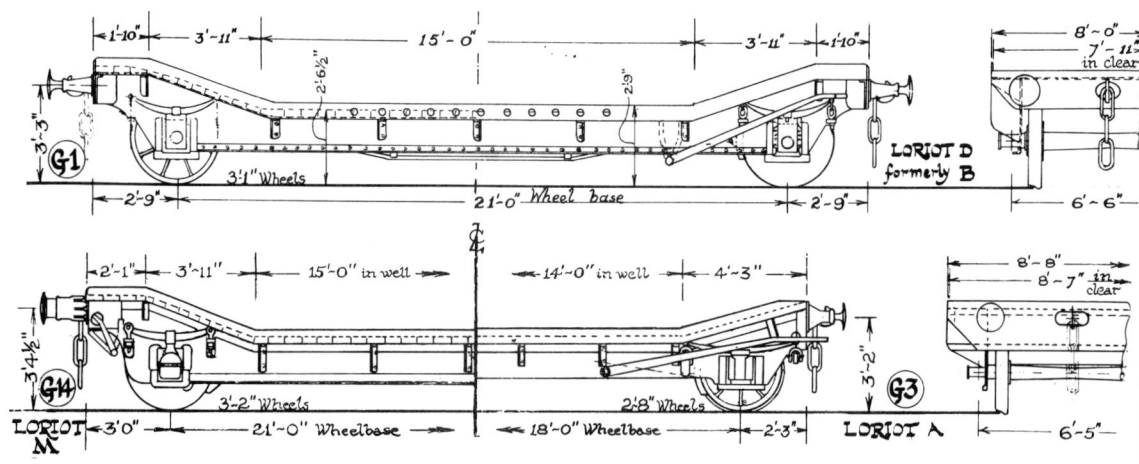

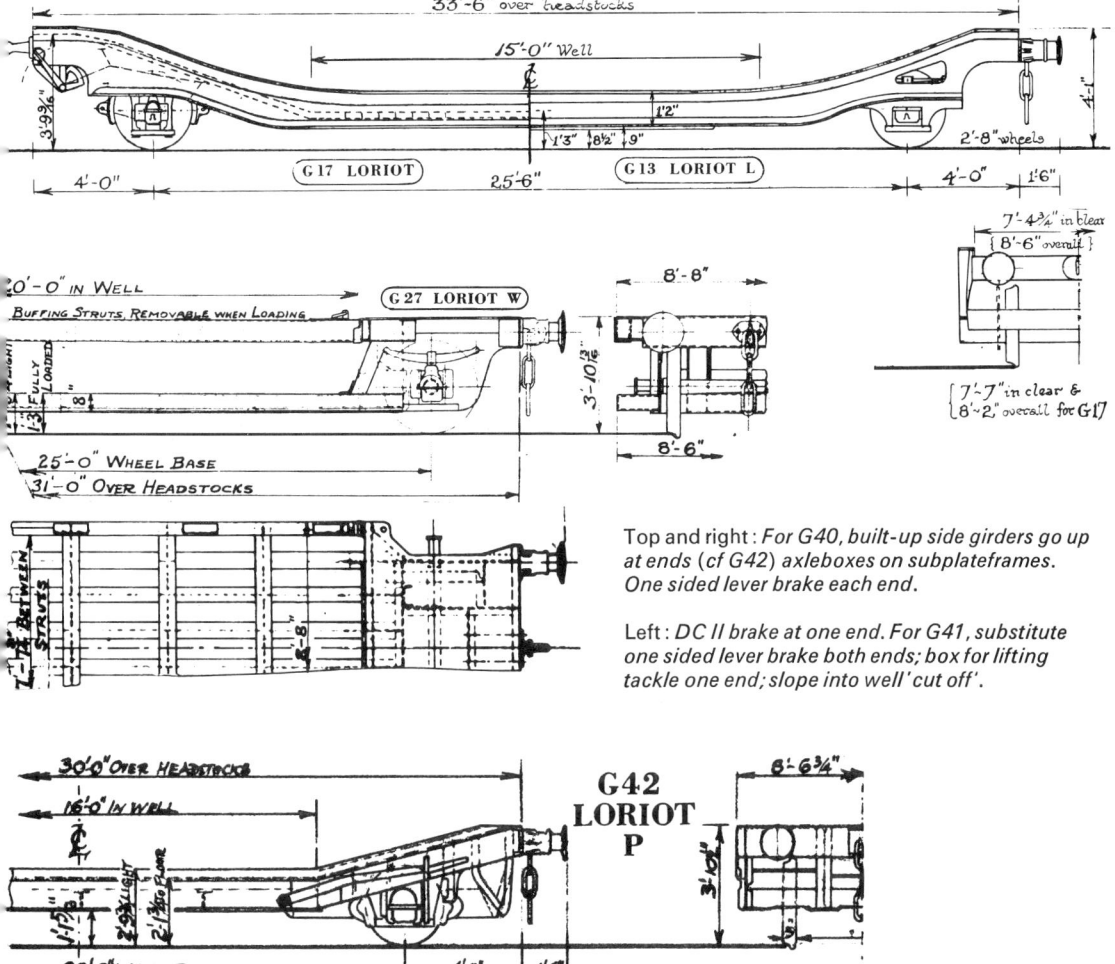

Top and right: For G40, built-up side girders go up at ends (cf G42) axleboxes on subplateframes. One sided lever brake each end.

Left: DC II brake at one end. For G41, substitute one sided lever brake both ends; box for lifting tackle one end; slope into well 'cut off'.

wheelbase, differences arising in width, buffing and brakegear. The early codes were altered about 1909. The service LORIOT P of 1943 reverted to earlier concepts, including the use of 2ft 8in wheels as G3/4, omitting end plaftorms and having long ramps to a 16ft well.

What may be called 'curvy' LORIOTS formed a slightly different well-wagon design. The profiles of G17/13/40 were similar (but not identical) in that the ramp side-frames blended into the well in a continuous curve; all had 25ft 6in wheelbase and 2ft 8in wheels. One such six-ton vehicle was introduced in 1905, for 'portable' engines (stationary traction engines) coded LORIOT, and used the same number (31308) as an earlier condemned 'flat-sided' LORIOT; the surviving 'flat-sided' LORIOT (31307) was then recoded SERPENT, as mentioned above. G17 was important from a design point of view, as the concept of continuous built-up side girders was introduced, which four

years later changed the construction methods of bogie CROCODILES (type VI). Fifteen vehicles of G13 followed in 1927-31 as the 15-ton LORIOT L, and 35 wagons of the 20-ton G40 LORIOT N in 1940-4.

The reason why G13 was a newer design than G17 was that before the grouping, diagram numbers were not re-used upon withdrawal of a design; the question arose on few occasions anyway (cf C8 CROCODILE J). For example, the 1880 HYDRA A 5878 had been scrapped soon after the index was set up, thus vacating the G15 diagram. Nevertheless at least G21-3 were all issued between 1908-20 as new diagrams after its withdrawal. However, with the amalgamation, G15 was given to the TVR well-wagon coded LORIOT F. Although diagram numbers were not issued for the absorbed LORIOTS G/H, the Rhymney LORIOT J took G5 that had been vacated in World War I when the SERPENTS A/B became

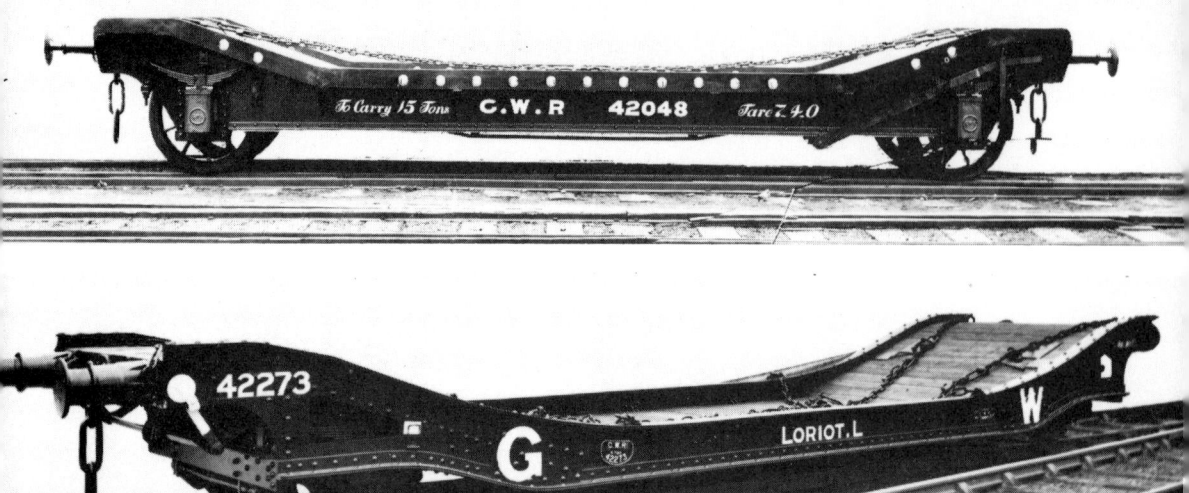

Top: *G1 LORIOT B (old code; D after 1909) built 1890, photographed about same time. D-shackles 3ft springs. Grease boxes. Simple one-sided lever brake. Holes for lashing. Underframe trussing (RWA Fig.252 for G2).*

Above: *G13 LORIOT L photographed when built 1933. Outside framing allowing 7ft 4¼in wide well. DC II single ended brake. GW self-contained buffers. 2ft 8in wheels. Works plate at slight angle.*

Below: *G39 LORIOT Y built and photographed 1939. Modern equivalent of C10 CROCODILE. Built up continuous I-beam girder slung outside axleboxes (cf width C10). 10 by 6in OK oilbox. Legend reads 'Where possible this vehicle must be marshalled in the rear of the Train and GREAT CARE TAKEN IN SHUNTING'. Self-contained buffers, large heads. Two separate DC II brake systems at each end.*

Bottom: *G42 LORIOT P photographed when built 1945. Frame riveted from I-beams with patch plates. Springs hung from large U-brackets attached under I-beams. 10 by 6in OK boxes, 2ft 8in wheels. Independent lever brakes both sides.*

roll wagons, and the Rhymney LORIOT K was given G8 when some very old SERPENTS from the 1870s were condemned. The trend having been set, the new GWR LORIOTS L/M took G13/14 in 1927, some old HYDRAS disappearing.

The LORIOT M of 1927 was the last well-wagon built with end platforms, and it was not until 1940 that new 'curvy' G40 LORIOTS N were built. A second lot of G40 appeared in 1943-4, but on the following L1448 the new G42 LORIOT P was ordered. This design abandoned built-up side girders, and used structural I-beams riveted with patch plates.

In the 1930s two odd LORIOT designs were produced in limited numbers. Intended principally for engineering excavators and the like, the two G27 LORIOTS W of 1931 were trolleys with very low, wide wells (BGW p.113). They had no ramps and resembled the T1/12 chaired sleeper wagons. Two more were built after the abandonment of DC brakes, so G41 with lever brakes was to G27 as T13 was to T12. Additionally a box for lifting tackle was provided on G41, (RWA Fig 254). Both diagrams had removable buffing struts along the sides of the deep well, but the wagons were in traffic without them; removable davits were also provided and sometimes used for loading. In later years, 100701 was used permanently to carry a trench digging machine. In 1937-9 two LORIOTS Y were built. At 32ft over headstocks, with 20ft well, they performed the same function as the LORIOTS W. Indeed, the lots called them

LORIOTS W originally, but this was amended, for they were built quite differently, using the design VI CROCODILE built-up girder principle (rather than rolled beams plus end axleguard plates as in G27) and really were modern wide versions of C10. Again, like the C group, two independent DC II brake systems were used at the ends. All the LORIOTS W/Y were marked 'Empty to Swindon'.

Vacuum pipes were fitted to some GWR LORIOTS (for example G2 42018/19/25/8/40 under L480 in 1906), and later instanter couplings. Some of the Welsh machinery trucks (ex-Rhymney LORIOTS J/K/H 42275-7 respectively) also had vacuum pipes and screw couplings, but were not intended to run in passenger trains.

Since the HYDRAS were intended to be marshalled in passenger trains, they were later classified as brown vehicles. When the index was set up, G10-15 were assigned to the extant HYDRAS in order of weight. The 10-ton G10 was the oldest design, built along with some of G3/4 under osL214 in 1880 (although with a 1ft deep wooden frame). Similar 2ft 8in wheels were used. Also from that year was the wooden-framed 3-ton G15 5878. Both were coded HYDRA A which, according to the 1900 *Telegraph Message Code Book*, meant *non-vacuum* well trucks to run in passenger trains. G15 was soon scrapped, but G10 was eventually rebuilt in 1925, downrated to 4 tons and recoded RODER. Undertrussed flat-sided four-ton HYDRAS, all similar to G15, were built with steadily increasing

The following were all 8ft 3in (7ft 1in) wide, and like G13 (orig), but for G15 (orig), 18ft oh, 11ft 6in wb, 8ft 6in well; for G14 (orig) 19ft oh, 14ft wb, 9ft well; for G12 (orig), 11ft well. Well floor-board edges showed on G12/14.

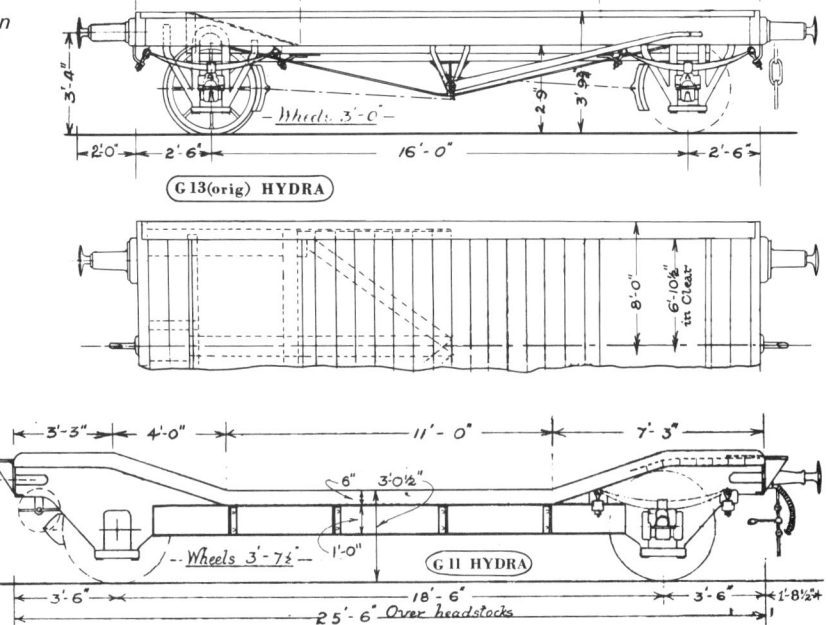

For G19, substitute 8 by 4in journals for 10 tons, DC II brake, coupled to single clasp-type blocks on far wheels as well as adjacent wheels.

lengths and wells between 1881-90, and became G14/13/12. One of these vehicles, G13 27276, was the first goods wagon to receive oil boxes.

It was not until G11 of 1899 (Vol.1, p.56) that obvious wells appeared in side profile. These vehicles had somewhat higher wells with more gradual ramps than G1 for example, because of the coach wheels, but since they were iron-framed they were much wider at 7ft 11¼in in clear than earlier HYDRAS (⅜in side rails to give 8ft overall). Larger 8in by 4in journals were provided on ten more similar vehicles rated at 8 rather than 6 tons in 1908; since also they had DC II brakes instead of Thomas, the new diagram G19 was issued. At the time of compilation of the index in 1904, two special wagons were constructed for the GWR motor department buses (G16). With much more gradual ramps, these 5-ton HYDRAS C were longer at 30ft 6in over headstocks, but had the same width as G11/19. One final passenger well truck was designed in 1913. Wagon wheels 3ft 2in in diameter allowed a longer well than G16 (15ft rather than 12ft 6in) with gradual ramps even though the over-headstocks length was shorter at 28ft 6in, and the 8ft 7¼in width was similar to G20 LORIOTS E. Screw couplings and 1ft 8½in

buffers were employed however on these ter HYDRAS D (G22). Incidentally, the C17 CROC-ODILE K built in 1909 for buses was longer and lower than any of the HYDRAS, but was only 6ft 7in wide in clear.

The HYDRA B code did not appear in the 1900 code book, yet the 1904 G16 wagons had the C suffix; the explanation for this is not known. Although later painted brown with ochre lettering, the HYDRAS did not change their running numbers upon reclassification.

SERPENTS were first given G5-9. Some 18ft over headstocks, 11ft 6in wb wood-framed 8- and 9-ton vehicles, which had guide 'ways' for agricul-tural machines and which dated from the 1870s, became G8; then 17ft 10-ton wagons followed as G7/6 from the 1880s, G7 being the former LORIOT. Vehicles 18ft over headstocks and of 11ft wheelbase were standardised in 1889, and by 1891 seventy had been built to the G9 design. Heavy six-wheel trucks were the SERPENTS A (15-ton) and B (20-ton) from 1877; these 'engine trucks' were noted in the lots to have had their ½in iron frames built at the wagon shops (rather than loco factory), and they were developments of the wooden H1 BEAVERS A. They were grouped

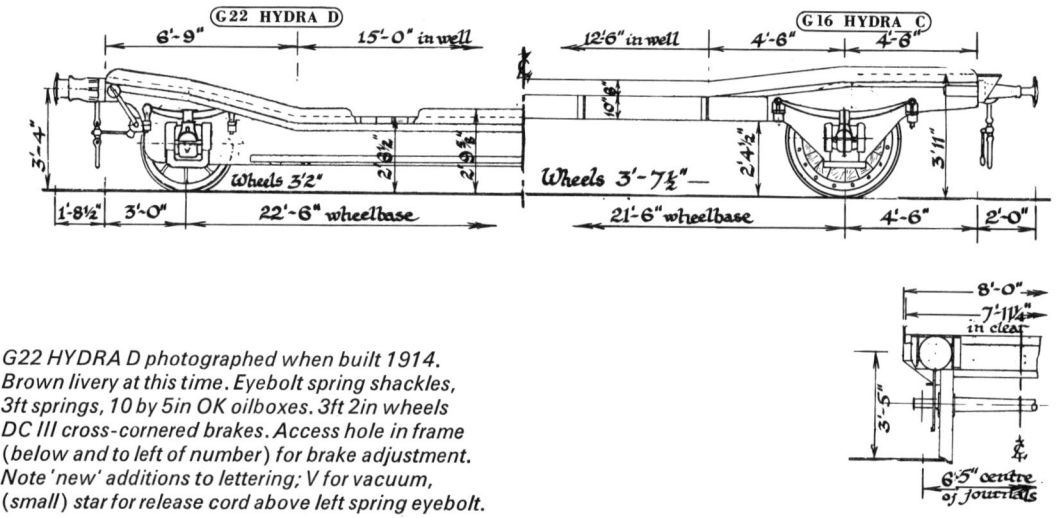

G22 HYDRA D photographed when built 1914.
Brown livery at this time. Eyebolt spring shackles,
3ft springs, 10 by 5in OK oilboxes. 3ft 2in wheels
DC III cross-cornered brakes. Access hole in frame
(below and to left of number) for brake adjustment.
Note 'new' additions to lettering; V for vacuum,
(small) star for release cord above left spring eyebolt.

G21 SERPENT C built 1913. Photographed 1930 after channels built up for conveyance of insulated containers. GWR 8 by 4in boxes. DC III vacuum brakes (central stripe) rod tiebar. Instanter couplings, GWR laminated sprung buffers (10½in covers as built for driving on vehicles).

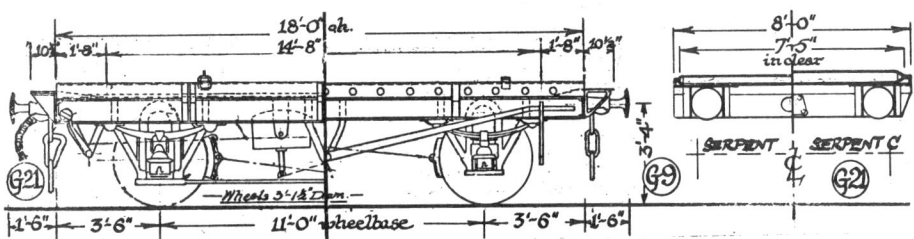

together into G5, not being separated until their conversion to roll wagons at the time of World War I. DC III vacuum versions of G9 were built between 1908 and 1913, and were coded SERPENT C. These G21 vehicles, like G9, had movable scotches that slid on the curb channels for retaining the load. All SERPENT wagons had very slight ramps at the ends (hidden by the curb rails) which prevented vehicles rolling off.

Only G9/21 survived after the grouping, most machinery being taken by LORIOTS, and most motor-car traffic going by covered vans (although open SCORPIONS were used in the Severn Tunnel car trains). The flat trucks built for container traffic were similar to the SERPENTS, and even before the building of the CONFLATS, order R.43 of May 1930 built the floors up level on 25 vehicles from G9/21 for E and FX insulated containers. Those G9 SERPENTS affected were subsequently recoded 'C' (42067/73/9/83/91/4/9/106). The G21 wagons reserved were 42122/7-9/33/7/65/7/8/71/5-8/82-4. See Chapter 7.

The World War II code reorganisation replaced the unfitted SERPENT code by CARTRUCK, and the fitted SERPENT C by CARFIT. The ex-TVR 10-ton 21ft 3in over headstocks SERPENTS D (42614-23, 42764/5), for which no diagram had been issued, were recoded CARTRUCK A. Since

there were no fitted goods carriage trucks longer than 20ft, the CARFIT A code did not apply to any GWR wagons.

One type of flat wagon was not incorporated in the G index until BR days. To deal with the portion of the Graig Ddu Slate Company's traffic shipped at Portmadoc over the narrow-gauge Festiniog Railway, the GWR provided wagons fitted with two sets of narrow-gauge rails upon which the slate trams were placed. They were taken in standard-gauge trains to Blaenau Ffestiniog, where the trams were run off on to a 1ft 11½in gauge line on a raised wharf. As many as six trams could be carried on the special GWR wagons, which were six-wheel ex-BEAVER C types 25020/9 (converted under L899 April 1923). In 1935 a four-wheel single V-hanger ex-MACAW A (48920) was converted as a replacement for 25020 (taking that number), but it was not until 1953 that this wagon was assigned G53.

Mayflys

New Work Orders F.404/5 in November and December 1919 converted the underframes of six condemned brakevans (old outside-framed 18ft type) into well-trucks for carrying electric transformers. An octagonal box well, 11in below the

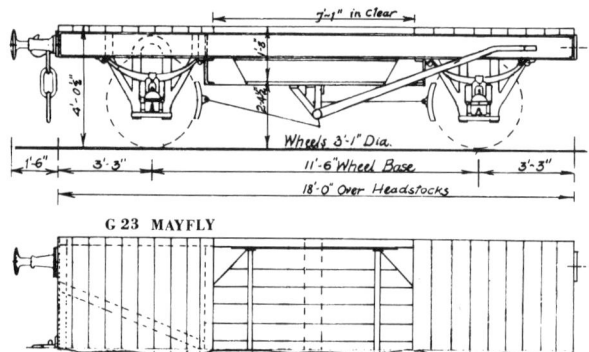

G 23 MAYFLY

Above: *G23 MAYFLY photographed 1920 when converted from AA-16 type brakevan. Old J-hangers, 4ft 6in springs, unusual spring stops. One-sided lever brake. Withdrawn 1940.*

Left: *Others converted from 18ft goods brakevans, with J-style hangers.*

Below: *For G25, shorten to 20ft oh, 12ft wb, modified u/f. For G26, ASMO, lengthen to 33ft oh, 22ft wb, angle trussing. For G32, as G26 with folding end doors and narrow ventilators. Note 4-V-hangers on long wb.*

solebars, was made between the wheels and frames, 7ft 1in long by 6ft 8in wide, the rest of the underframe being planked over. These G23 wagons were all condemned by September 1940 (RWA Fig243).

The much larger transformers that were built in the late 1920s and 1930s were catered for by the C24/5 CROCODILE L and C26/7.

Damos, Asmos, Bocars and Mogos

After World War I the number of motor cars in Great Britain increased from less than 200,000 in 1920 to over a million by 1930; there was a similar expansion in lorries and vans – 85,000 in 1915, 101,000 in 1920 and 350,000 in 1930. Although road competition attracted merchandise away from the railways, the GWR recognised that road vehicles could themselves be a source of traffic. Covered and open-carriage trucks had been part of the scene since the earliest days, for broughams, coaches and horses, and though this traffic declined after the war, the demand for the transport of cars increased. Hence in 1925 there appeared the first DAMO A and DAMO B (G24/5) vol.1, p.89, soon to be followed in 1930 by the ASMO (G26). All three were variations of the same basic vertically

planked, inside-framed design with end doors to allow cars to be driven on (the SIPHON J of 1930-1 was similar in appearance). Fifteen DAMOS A (30ft over headstocks for two cars) were built under L972 and ten DAMOS B (20ft over headstocks for one car) under L973. Although twenty more G24 were built in 1929-30, a longer 33ft over headstocks design was chosen to be built in quantity and 100 G26 ASMOS were constructed under L1059. Some had concertina-folding end doors that could be opened while the wagons were coupled together in a train, enabling cars to drive through; most however would be loaded separately at an end dock because of the arc of swing made by the normal double doors. The concertina door wagons, which were given the separate diagram G32, could be identified by the small end vents. All G24-6/32 were fitted to run in passenger trains, forty of the ASMOS receiving through steam pipes as an afterthought in December 1930 at a cost of £10 each. Instead of continuing to build special vehicles for motorcar traffic, convertible MOGOS were introduced in 1933-6 (G31). Essentially these were contemporary V23 17ft 6in RCH goods vans with end doors (Vol.1, p.82), the car chocks and holding-down

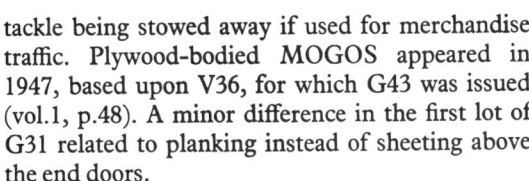

tackle being stowed away if used for merchandise traffic. Plywood-bodied MOGOS appeared in 1947, based upon V36, for which G43 was issued (vol.1, p.48). A minor difference in the first lot of G31 related to planking instead of sheeting above the end doors.

Regular traffic in motorcar bodies built up between Oxford and the Midlands in the late 1920s. As a trial, the GWR built a superstructure on to MACAW E FLAT 70736 in 1927 that would accommodate nine finished Wolseley bodies mounted sideways, to be taken to Coventry for chassis. Protection from the weather was given by tarpaulin sheets. This was the forerunner of the forty-six BOCARS, the diagrams for which (G28-30) were formally issued in 1930 and 1933. All were converted MACAWS B (J14/21) or E (J23/4), with detail differences in design. For example, NWO R.48 of October 1928 converted 29 MACAWS B (G29) and Order 9/600 in 1933 modified six MACAWS B (G30). One of G28 (70739) was fitted with end doors. Although successful in concept, it was extravagant to use

Above: G32 ASMO built 1930, photographed with small GW lettering in 1937. Special concertina folding end doors, note hinge straps and small end vents. Rectangular works number plate. Large heads on self-contained buffers. Lamp irons. Destination label clip on body not solebar.

Above left: G28 BOCAR photographed when converted in 1927 from MACAW E FLAT J24. Latter code left on wagon. Originally overseas military MACAW D J19 of 1917. 5ft 6in plateframe bogie, external stiffeners, 8 by 4½in 'square' boxes. Large head GW self-contained buffers. Swivel coupling hook. Double card frame on label box. Note steel plate floor on battens. Straps on ends for lacing tarpaulins. Reconverted to J24 FLAT in World War II.

MACAWS B/E rated at 30 tons to carry 5 tons of car bodies, so the fleet of 46 BOCARS was augmented in 1934-8 by vehicles with similar bodies mounted on old carriage underframes. There were 136 bogie carriages thus converted before the war (G33/7/8) coded BOCAR A, and 24 four-wheel coach underframes G34/5/6 coded BOCAR B. Differences between the diagrams arose with lengths and wheelbases depending on the

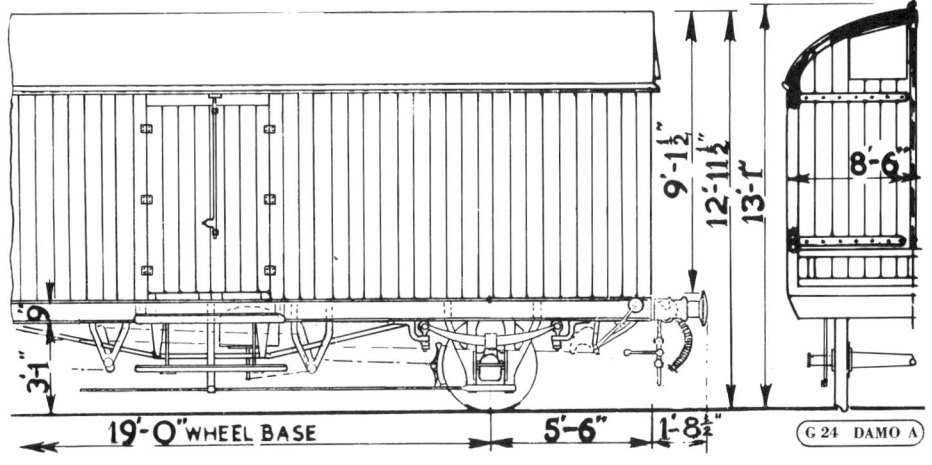

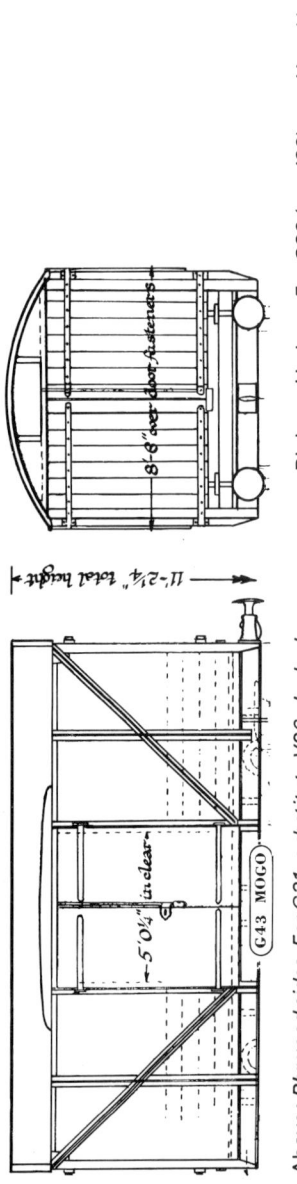

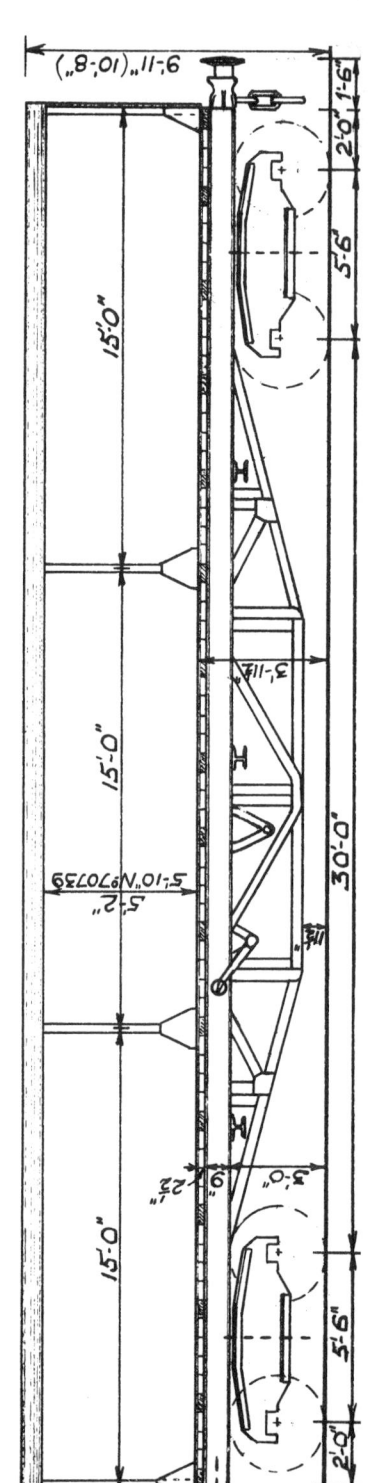

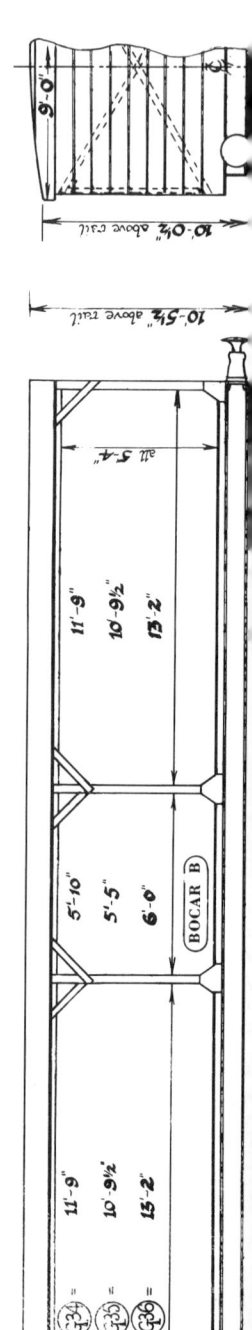

G 28 BOCAR

8'8"

4'5½"

5"

5"

9'6 (10'3" 70739)

3'5"

8'8" over door fasteners

G43 MOGO

11'2¼" total height

5'0½" in clear

Above: Plywood sides. For G31, substitute V26 planked body and older u/f (no longitudinal scantlings).

Right and below: For G29 (ex-J23), oval head buffers, four panels at 11ft 3in centres; X-bracing on ends. For G30 (ex-J21), as G29 but s/c circular buffers; for G30 (ex-J14) (84643), as G29 laminated buffing and drawgear.

Bottom: G34, 30ft 5in oh, 19ft wb; G35 27ft 5in oh, 18ft wb; G36 33ft 5in oh, 22ft oh.

9'11" (10'8")

2'0" 1'6"

5'6"

15'0"

15'0"

30'0"

3'11"

5'10" N°70739

5'2"

15'0"

5'6"

2'0"

9'2"

2'2½"

5'0"

9'0"

10'0½" above c.sill

10'5½" above rail

5'4"

11'9"

10'9½"

13'2"

BOCAR B

5'10"

5'5"

6'0"

11'9"

10'9½"

13'2"

G34

G35

G36

Top: *G33 BOCAR A photographed when converted in 1934 from 6ft 4in Dean corner-hung bogie carriage underframes. Note removable plank 'barriers' on far inside. Black u/f, grey superstructure.'To carry' out of style for period.*

Above: *G34 BOCAR B converted in 1934 from 4-wheel carriage underframes. Photographed 1944 when modified for 'special traffics'. Canvas sheeting permanently tacked down and doors added, 'to carry' out of style. Legend is 'To work between Paddington, Swindon and Clifton Bridge only'.*

original carriage. The vehicles retained their carriage wheels (and 6ft 4in or 8ft 6in Dean centreless scroll iron bogies) and all received centrally positioned DC II brakes (RWA Fig 215).

The origin of many of the GWR telegraph code names is enigmatic, but here there appear to be connections between the vehicle and its traffic. ASMO is derived from ASsembled MOtor cars, MOGO from MOtorcar GOods (train), and BOCAR from BOdy motor CAR. This pretty idea breaks down however with DAMO, since they were not DisAssembled MOtorcars!

With the coming of World War II, car manu-facture ran down and more particularly the need arose for heavy bolster wagons for military material. Consequently, G28/9/30 were reconverted to their original form, G30 being finally completed by the beginning of 1941. A further change to FLAT wagons in 1943 is dealt with in the J section. Necessity being the mother of invention, nine-ton rail wagons were also made from some of the carriage underframes G33/7/8 in 1940, coded MACAW Z (BORAIL A after 1943). That these wagons were not typical is seen from the coding (cf LORIOTS R, W and Y). Although they were a useful addition in time of need, difficulties were encountered in negotiating curves in the South Wales docks, four chain radius curves being the minimum traversable. After the war they were reconverted to BOCARS A (except 107417, which had been converted in 1941 to an engineering piledriving tender and renumbered 14026). Also in 1946 'new' conversions were made of other carriage underframes into BOCARS A (G44/5), a practice carried on into BR days for the Pressed Steel Company's traffic.

CHAPTER 7
H – FLAT WAGONS
(Beaver, Gadfly, Conflat)

H1 and H2 were the six-wheel BEAVERS A and the BEAVER B respectively. They were heavy 20-ton flat trucks with no bolsters. In fact the BEAVERS A, dating from 1871, were the GWR's first really heavy-load special wagons. Four were built under osL47, 13619/20 being 17ft 6in over headstocks by 8ft with 7in deep sides and 13621/2 being 18ft by 7ft 7in. Slightly different lengths and widths were given for these wooden-framed wagons in the early issues of the *Special Wagon Book;* curiously the 1889 issue omits 13619/20. H2 dates from 1886, and 42901 is a much longer and wider metal-framed vehicle at 32ft over headstocks by 8ft 8in.

One of H1 was condemned in 1896, another in 1904, and 13619 followed in August 1910. The BEAVER E 41996 of 1911 was built as a direct replacement and given H3. Likewise 41995 of 1913 replaced 13621 scrapped in 1912 (cf the serial

numbers used for MORELS). The surviving BEAVERS A had been in use therefore for some forty years. A Swindon drawing of the proposed 20-ton 20ft over headstocks by 8ft 9in BEAVER shows stanchions, but they were not incorporated when built. Eventually H3 vehicles were down-rated to 15 tons and converted to roll wagons (B1C in 1942; their function had been usurped by MACAWS H and the like.

The diagram index states that the H group were flats for 'aeroplane traffic'; during World War many old carriage underframes were given plain decking with no curbs for conveyance of partially dismantled fuselages and wings. The carriage flats were the GADFLY wagons which occupied diagrams H4/5. H4 were two vacuum-fitted four-wheel ex-Rhondda & Swansea Bay Railway thirds and amongst H5 was included the 1889 BEAVER C (J6) which had its bolsters removed and was

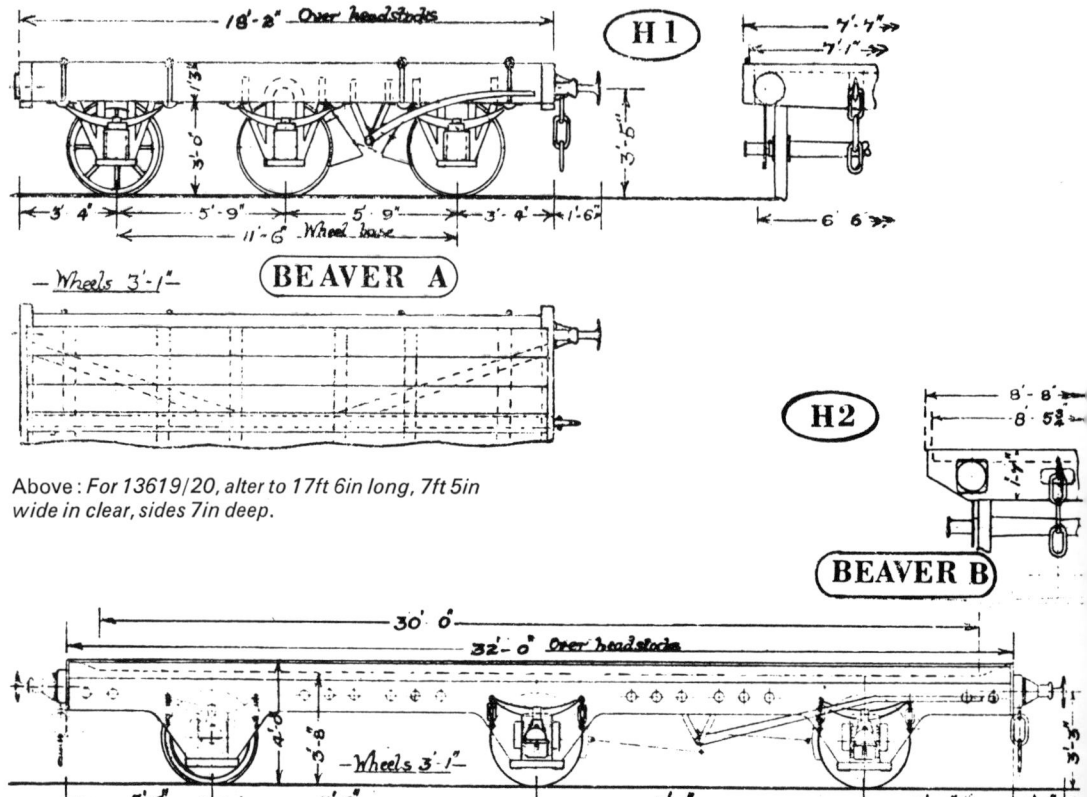

Above: *For 13619/20, alter to 17ft 6in long, 7ft 5in wide in clear, sides 7in deep.*

e-numbered 94679, but most of H5 were old wooden-framed carriages. Paradoxically most of the GADFLYS had bolsters and stanchions added after the summer of 1923 (order F.452).

The ultimate use of the H-group however was for container trucks. Open wagons and match trucks were used to carry containers in the early days (e.g. RWA Figs 169/76), but special CON-

FLATS appeared in the 1930s, some G9/21 SERPENTS (C) having been altered for container traffic in 1930. The CONFLATS were contemporary OPEN underframes with 4¼in raves fitted all round. Thus H6 has the (9ft + 4ft 3in) underframe of O29, H7/9/10 the (10ft + 3ft 9in) underframe of O32 and so on. The twelve wagons of H8 were converted from the 18ft over headstocks

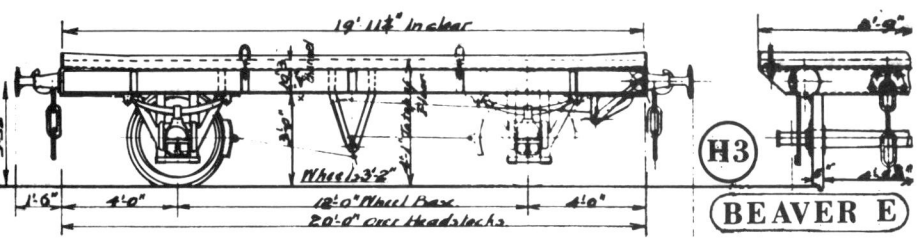

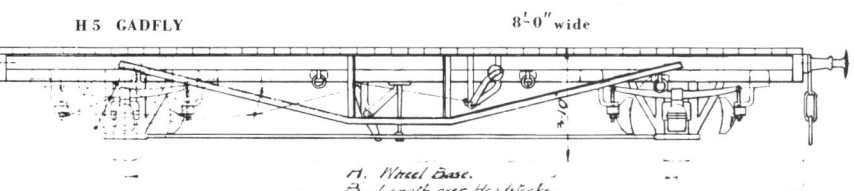

Wagon	"A"		"B"		Tare	
94676	15ft	6in	27ft	0in	6 tons	7cwt
77	15	6	27	0	6	6
78	19	0	27	10	6	8
79	18	0	27	7	6	9
81	15	6	27	0	6	6
82	15	6	27	0	6	6
83	19	0	27	10	6	8
84	19	0	29	10	6	17
90	18	4	27	10	6	14
91	19	0	28	10	6	5
92	19	0	27	10	6	6
93	19	0	27	9	6	9
94	19	0	27	10	6	6
95	19	0	27	9	6	6
96	19	0	28	10	6	10
97	19	0	28	10	6	8
98	16	0	24	10	6	7

For H4, J-spring hangers, vacuum brake.

Note some GADFLY running numbers in above list were omitted in Chapter 3, Table 7.

Above: *H5 GADFLY with table of dimensional variations, left.*

H5 GADFLY photographed when converted in 1917 from J6 BEAVER C (built 1889). GW oilboxes. Instanter coupling, but not Gedge hook. Other GADFLY wagons converted from wooden carriage u/f, vacuum fitted with underslung J-hanger suspension and trussing.

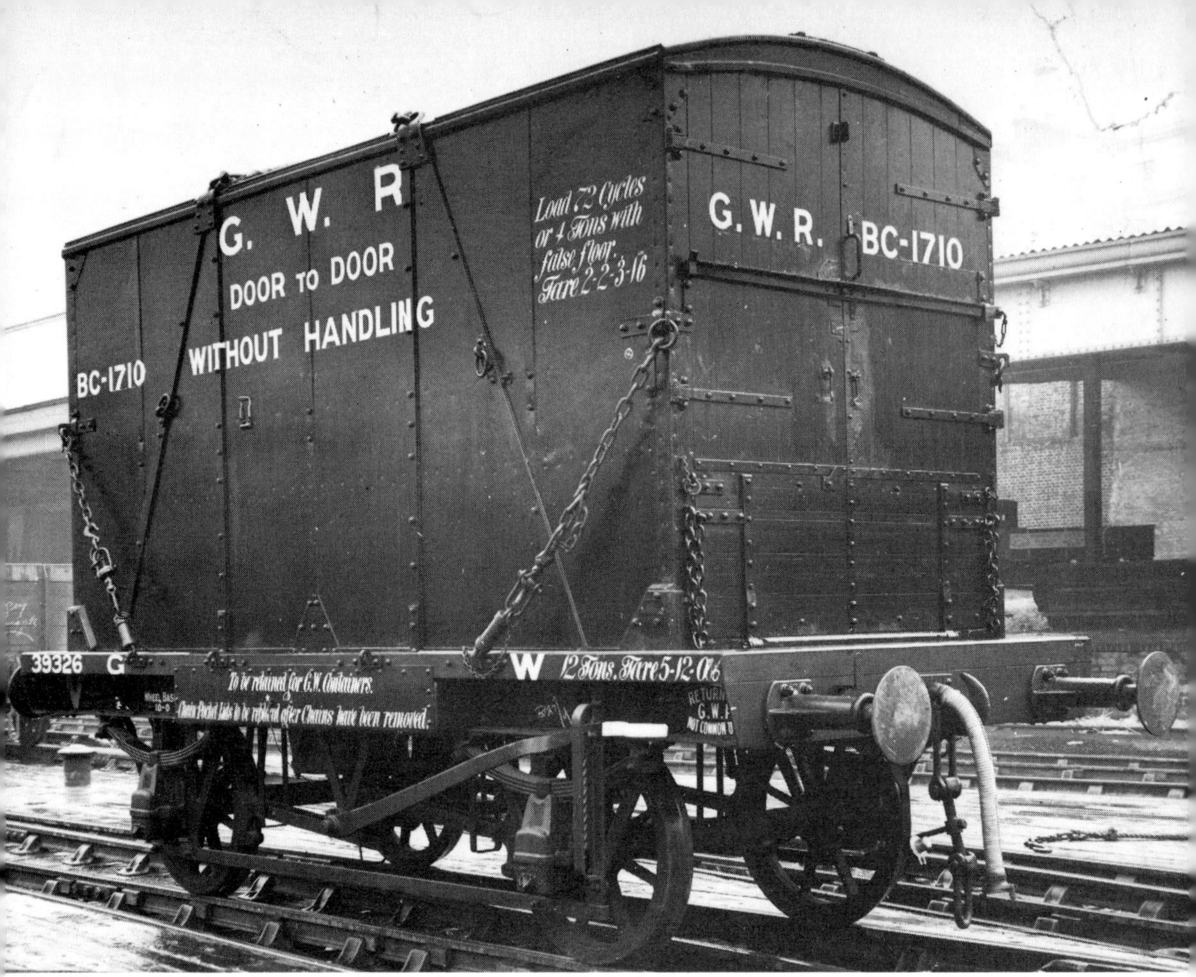

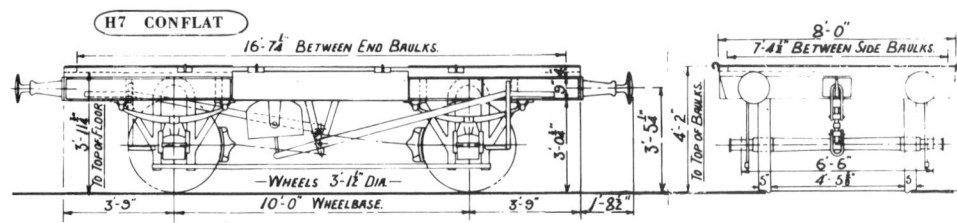

H7 CONFLAT

Top: *H7 CONFLAT built 1933, photographed with bicycle container. RCH 17ft 6in vacuum u/f, flat strip tiebars. Long rib RCH buffers.*

Above: *For H6, substitute 9ft wb u/f (as V21), chain pockets on ends. For H8 substitute 18ft oh, 10ft 6in wb, ex-milk tank u/f, J-hangers. For H9, as H7 but non-vac, 3-link couplings, no tie bars. For H10, as H7 with extra longitudinal u/f scantlings (externally same as H7). All 16ft 7¼in between end baulks.*

frames of milk tank trucks that themselves had not been altered to six-wheels; as such they had underslung springs. Some were later converted to demountable varnish tank wagons (EE1). Chain

pockets below the underframes at the ends were provided on H6 (RWA Figs 161/5/8/74), but were absent on the remainder (RWA Figs 162/4/6 (twice) /7 – most with disc wheels). Four eyes were provided on the sides of the wagons (and also two on each end of H6) for attachment of chains and shackles. The 6ft by 1ft plates on the solebars had the legend in script 'To be retained for GW Containers. Chain Pocket Lids to be replaced after Chains have been removed.'

All container vehicles, except for the 200 wagons of H9 produced in 1943-4, were vacuum-fitted and provided with lamp-irons. The 'A' subscript on CONFLAT was intended to be used on H9, but as their introduction coincided with the war standardisation of the telegraph codes (which reversed the meanings), H9 became CONFLATS and all the others CONFLATS A.

CHAPTER 8
J – RAIL AND TIMBER BOLSTER WAGONS
(Macaws and Ganes)

There were no bogie bolster vehicles for carrying rails and baulks of timber until the early 1890s, when the two 33ft over headstocks BEAVERS D were introduced. Although there was the one 15 ton BEAVER C (25ft over headstocks, later rebuilt to 27ft) before this, long comparatively light loads would be taken by a series of 15ft MACAWS coupled together, or by the permanently articulated pairs of MACAWS called MITES and MITES B (31ft 1in over headstocks). Such four-wheel bolster vehicles, with detail differences, existed from the 1870s. Short heavy loads were the business of the other BEAVERS (in the H group), and long heavy loads that of POLLENS (A group) and TOTEMS (B group).

In 1899 a new 'special wagon to carry long rails' was introduced for the construction department. Coded GANE, it was 45ft over headstocks, and had massive 1ft channel underframe for its 40 ton load; the bogies were 6ft wheelbase sprung from the large exterior trunnions. Its running number (40997) had been allocated originally to a cancelled departmental steam roller wagon (F2), RWA Fig 218. Twenty further vehicles were built in 1900, numbered 40577-97 (not 40557-97 which is a typographical error in Vol.1).

A two-bolster version of an 1897 engineers' points-and-crossings wagon (T3) appeared in 1902. The two-plank 25ft 6in body was essentially similar, but the wheelbase was shortened from 18ft to 17ft. These became MACAWS A. Twenty were built for general traffic in the first lot, ten wagons (33951-60) for conveying loco-dept materials to out-stations in 1904, and one vehicle (14430) for the signal department replacing the wagon of the same number condemned in 1908.

A one-off 36ft three-bolster bogie wagon had been built for the signal department in 1903, but the following year saw the introduction of a design that was perpetuated for the next fifty years; the

J1 GANE built 1900 photographed 1910. 6ft wb plateframe bogies suspended from trunnions. 10 by 6in boxes. Thomas brake (horizontal OFF/ON plate) acting on both bogies. Flatstock trussing, plain kingposts. Round guide oval buffers. Despite plate 'FOR NEW RAILS ONLY', wagon fitted with cradles for carrying cast iron pipes to Swindon from Chesterfield, hence heavier tare.

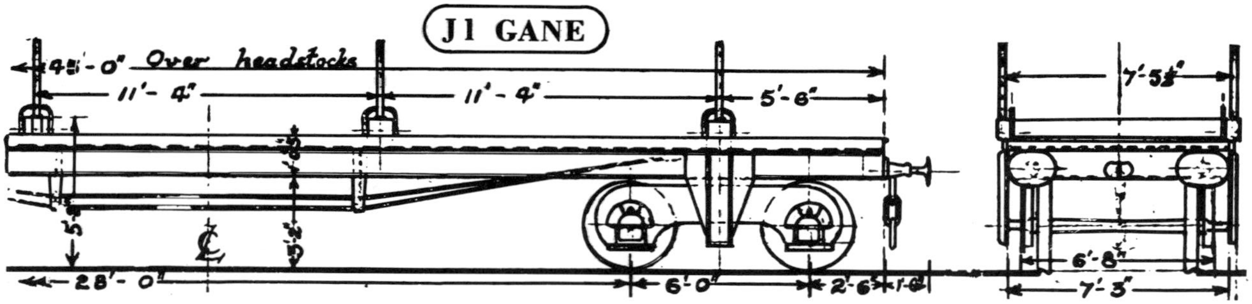

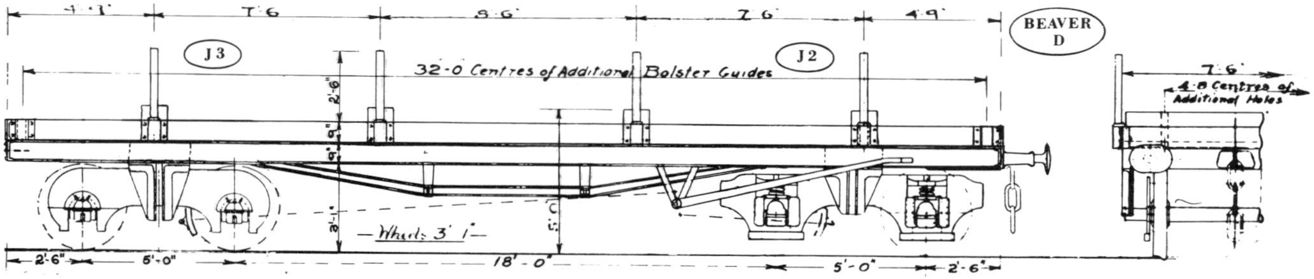

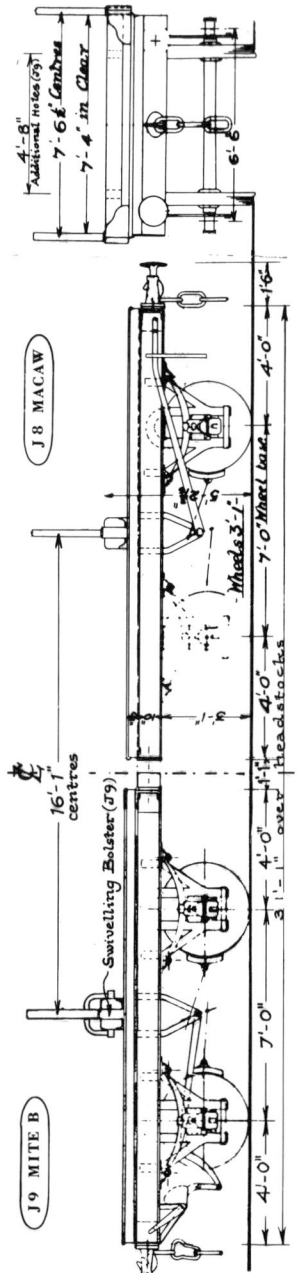

Below: For J8, single wagon (buffers both ends) as on right, fixed bolster. For variations on these drawings, eg wooden sides etc see photographs and text.

MACAW B had the same 30 ton capacity as the BEAVER D, but had the 45ft length of the GANE. Significantly the vehicles were equipped with the modern 5ft 6in plateframe bogie that was then being fitted to the MINKS F, CROCO-DILES and loco coal wagons. The first lots were built with the bogie centreline 5ft 3in in from the ends and a single-ended DC II brake; after 1907 the bolster positions were altered, the wagon was widened, and the bogies were moved outwards so that the end axles were only 2ft from the head-stocks.* A similar change took place with the bogie coal wagons, but the original 2ft 6in dimension remained on all the bogie CROCODILES.

The setting up of the J diagram index was a little illogical. Certainly GANES were J1, but following there should have been the MACAWS B; instead came the BEAVERS D as J2/3. Two diagrams were issued because of the curious coil spring bogie on J2. Then came the MACAW B, except that it was the 1907 design. Even if the J

*The 1910 *Special Wagon Book* erroneously gives the total wheelbase of numbers 70807-910 (J11) as 41ft; it should have been 40ft.

group was set up after 1905, so that the 1907 MACAW B was the newest, the original 1904 design should have been J5, instead of which it was J11, tacked on as an afterthought. J5 actually was the signal department bogie bolster, followed in descending capacity by the BEAVERS C, MACAWS A, and MACAWS. J9 was the MITES and MITES B, which should have been inter-changed with the J8 MACAWS. Finally came some engineering department service vehicles, converted from old broad-gauge wagons akin to the BEAVERS C, which again were out of sequence. Then the 1904 MACAW B design was put in.

The first new vehicles added to the list (J12) were the 1909 14-ton rail wagons with fall-down sides and ends, which were used for switches and crossings; these could have been at home in the T-classification. They were fitted with a special loose flat instanter coupling which allowed them to to be coupled very closely together to share long loads. In 1913 L730 built four vehicles (numbered 95/155/61/9) for the Port Talbot Railway, the

Below: *J9 MITE B built 1881 (32243/4) on right and 1896 (48427/8) on left. Photographed 1912. Two different versions of same articulated design. On both, bolsters swivel; on newer pair bolsters ride over shallow steel curb rail; on older pair wooden planking cut away to allow swing on curves. All have D-chain shackles (therefore B code). Lettering is on planking of older set, but number and GW on solebar of newer. Buffers on older pair unusual, and one-sided brake has old curved lever. Swing arms rigidly attached to brakeblocks. Both have iron frames, older with horsehooks, newer with holes in solebars. By time of photograph, oil boxes fitted.*

Bottom: *J12 built and photographed 1915. J-hangers with 4ft 6in underslung springs. OK boxes. Square shank oval buffers. Flat strip trussing with plain kingposts. DC III cross-cornered brakes (central white stripe on solebar). Brake handle not painted. Fall-down sides and ends (removable side stanchions). Special instanter coupling allowed wagons to be close coupled with ends down to carry long sleepers or rails. Legend reads 'This Vehicle must not be close coupled on any Curve Less than 3 Chains Radius. Return empty to Hayes Creosoting Works'.*

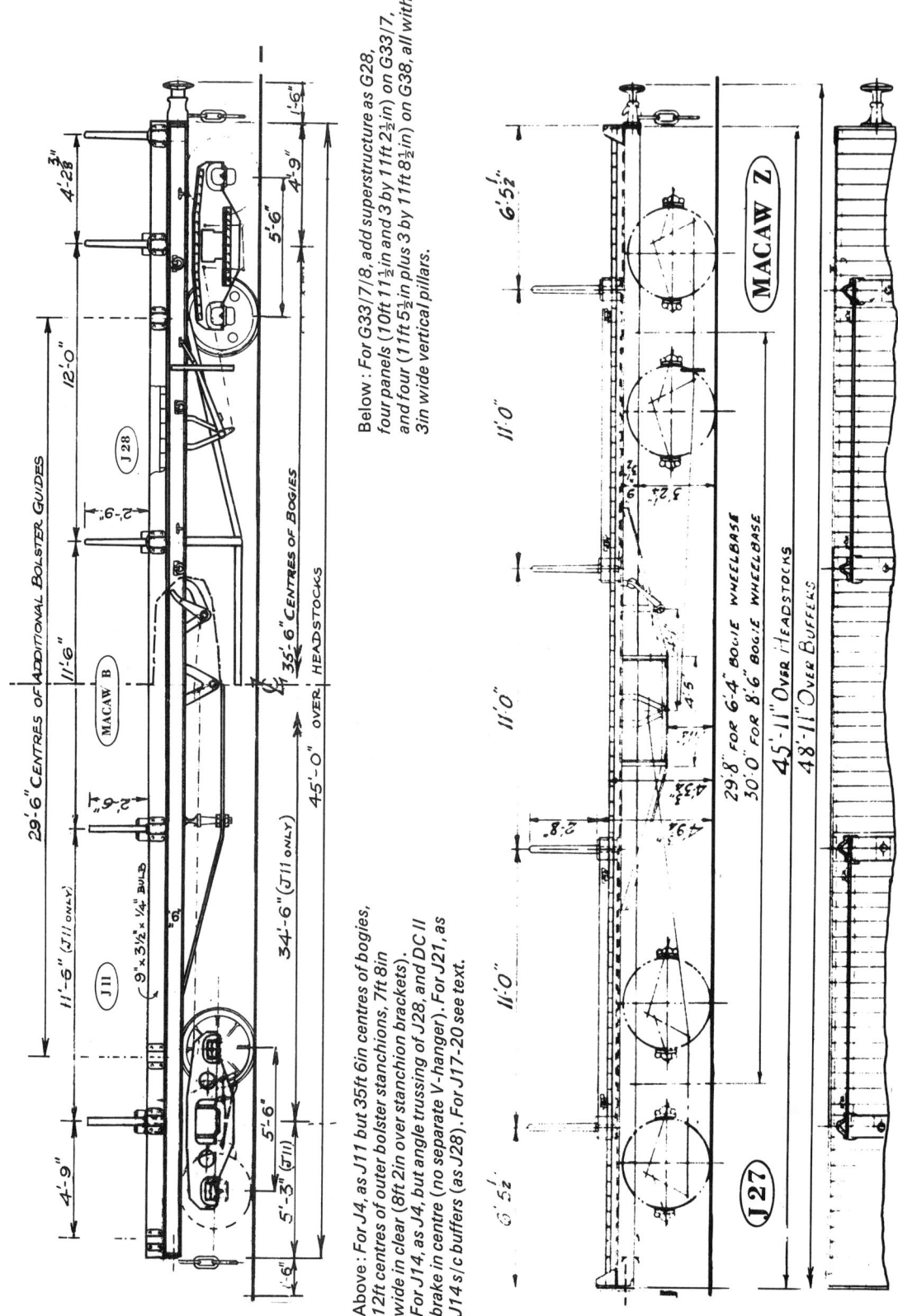

29'-6" CENTRES OF ADDITIONAL BOLSTER GUIDES

12'-0"

35'-6" CENTRES OF BOGIES

45'-0" OVER HEADSTOCKS

J11

MACAW B

J 28

11'-6"

2'-9"

2'-6"

9'

9" x 3½ x ¼ BULB

11'-6" (J11 ONLY)

34'-6" (J11 ONLY)

45'-0" (J11)

4'-9"

5'-6"

5'-3" (J11)

'-6"

4'-2⅜"

5'-6"

4'-9"

'-6"

Above: For J4, as J11 but 35ft 6in centres of bogies, 12ft centres of outer bolster stanchions, 7ft 8in wide in clear (8ft 2in over stanchion brackets). For J14, as J4, but angle trussing of J28, and DC11 brake in centre (no separate V-hanger). For J21, as J14 s/c buffers (as J28). For J17-20 see text.

Below: For G33/7/8, add superstructure as G28, four panels (10ft 11½in and 3 by 11ft 2½in) on G33/7, and four (11ft 5½in plus 3 by 11ft 8½in) on G38, all with 3in wide vertical pillars.

MACAW Z

J 27

11'-0"

11'-0"

11'-0"

6'-5½"

6'-5½"

29'-8" FOR 6'4" BOGIE WHEELBASE
30'-0" FOR 8'6" BOGIE WHEELBASE
45'-11" OVER HEADSTOCKS
48'-11" OVER BUFFERS

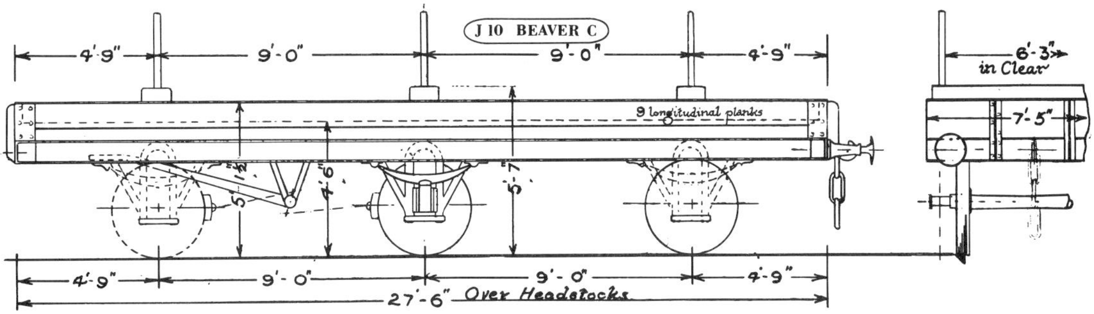

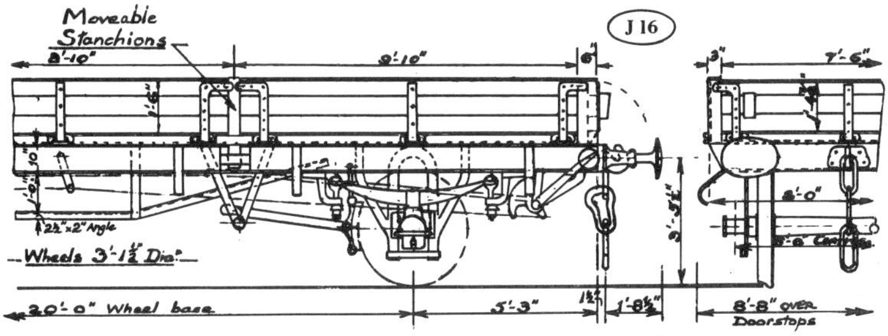

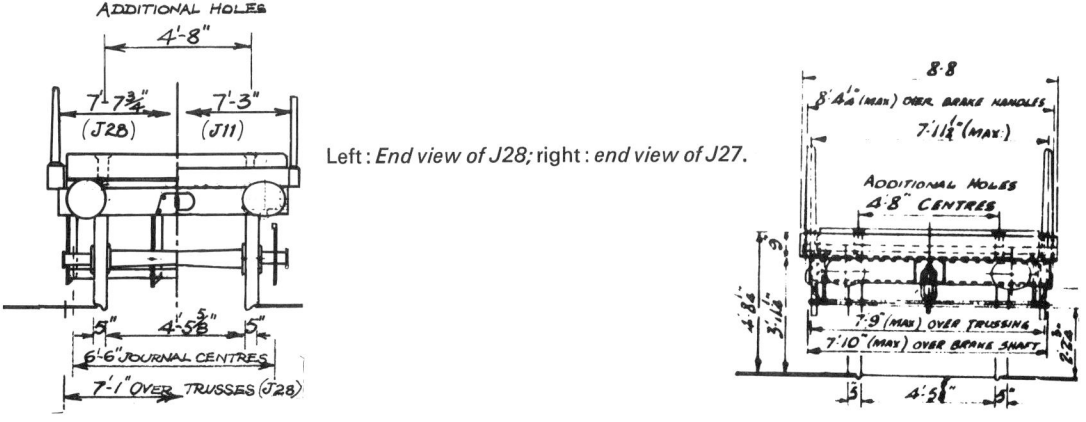

J 10 BEAVER C

4'-9" 9'-0" 9'-0" 4'-9"

9 longitudinal planks

6'-3" in Clear

7'-5"

4'-9" 9'-0" 9'-0" 4'-9"

27'-6" Over Headstocks

J 16

Moveable Stanchions
8'-10"

9'-10"

7'-6"

Wheels 3'-1½" Dia.

20'-0" Wheel base

5'-3"   1'-8½"

8'-8" Over Doorstops

Left: *For J12, substitute rod trussing, fixed kingposts. For L10/18, substitute 3ft 6in underslung springs instead of 4ft 6in, and adjustable kingposts; on body, 11 in top plank to give 2ft height and 5 equal side panels instead of 3; ends not fall down.*

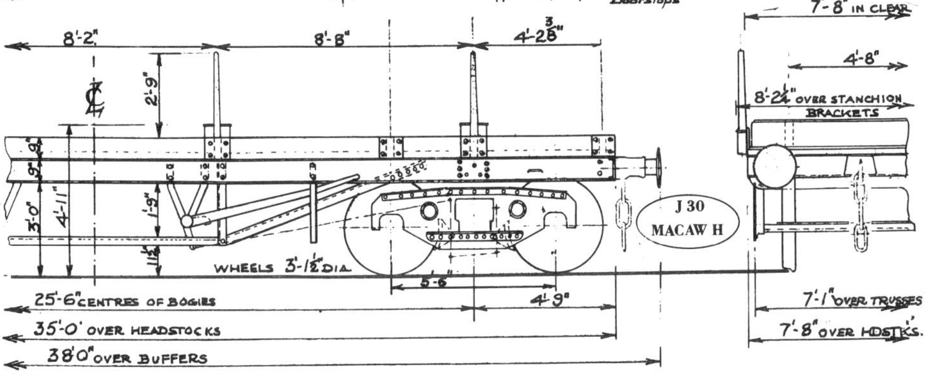

8'-2" 8'-8" 4'-2⅜"

2'-9"

7'-8" IN CLEAR

4'-8"

8'-2¼" OVER STANCHION BRACKETS

WHEELS 3'-1½" DIA.

J 30 MACAW H

5'-6"   4'-9"

7'-1" OVER TRUSSES

25'-6" CENTRES OF BOGIES

35'-0" OVER HEADSTOCKS

38'-0" OVER BUFFERS

7'-8" OVER HDSTKS.

For J25, substitute DC brake centrally (as on J11), add end curb rails.

ADDITIONAL HOLES
4'-8"

7'-7¾" (J28)   7'-3" (J11)

5" 4'-5⅝" 5"

6'-6" JOURNAL CENTRES

7'-1" OVER TRUSSES (J28)

Left: *End view of J28;* right: *end view of J27.*

8-8

8'-4¼" (MAX) OVER BRAKE HANDLES

7'-11¼" (MAX)

ADDITIONAL HOLES
4'-8" CENTRES

7'-9" (MAX) OVER TRUSSING

7'-10" (MAX) OVER BRAKE SHAFT

4'-5⅝"

Top : *J18 modified military MACAW B. Converted in December 1916 from J17 which in turn had been modified from J14 built earlier in 1916. Extra angle trussing, but still rated at 30 tons. Loading ramps. 5ft 6in wb plateframe bogies with external stiffeners. DC II brake. Square shank oval buffers. Legend says 'For Military Purposes only. Return to Swindon when not in use.'*

Above : *J26 GANE A photographed when built 1938. 62ft oh, 10in solebars, 51ft 6in bogie centres, 40 ton, 10ft centres of stanchions. 10 by 5in boxes exterior ribs top and bottom of plateframe bogie, disc wheels. GWR 1ft 7in head s/c buffers, but RCH drawgear. Two independent DC II brake systems. Oval works plate. Red stripe down curb rails. MACAW J is same, without end curb rails. J29 same, but with right-hand lever brakes either end, as shown on J28 vol.1, p.80.*

Bottom : *J31 BOGIE BOLSTER A photographed when converted in 1947 from RECTANKS C21, WD diamond-frame bogies, screw brake handle one end (cf old Thomas brake), self-contained buffers. RECTANKS 17xxx two extra lashing rings.*

Below : *J27 MACAW Z photographed when converted in 1940 from BOCAR A G33. Far end of solebar contains similar legend to BOCAR A 'To carry 5 tons distributed etc.' Later conversions had extra angle trussing under centre of solebars.*

working responsibility for which had been taken over by the GWR in 1906. Breakdown vans and other wagons for the Port Talbot Railway were incorporated in the Swindon lots at this time.

More 40-ton 45ft GANES were built in 1913 (J13) and were essentially uprated J4 with extra flatstock trussing; they differed markedly therefore from the original J1 GANES. The same year saw the introduction of angle trussing on the MACAWS B (J14), and thereafter further diagrams for MA-CAWS B related to differences in buffing and drawgear, brakes and the absence of end curb rails. Similarly J16 was an angle-trussed J12. The new trussing appeared on the 70ft MACAW C (J15), also introduced in 1913 (cf length of Dreadnought carriages).

Many MACAWS B were altered for the conveyance of guns, coastal motor boats and particularly tanks in World War I. The 41 vehicles of J17 were essentially wagons from J14 (including two from J4) mostly with the bolsters and stanchions removed (RWA Fig 224 shows stanchions in place). Fourteen of these were reserved for tracked-vehicle traffic between Avonmouth and Portsmouth. J18, called 'modified' MACAW B, was introduced in February 1917 on Order O.670 for carrying tanks in England, with underframes strengthened by two extra cross-braced rows of trussing under the middle of the wagons. Some were modified from J17, but the rest were conversions of other J14 vehicles. Unlike J17, the side raves were removed completely, thus exposing the floorboard edges; they also had long loading ramps, but were still rated at 30 tons (RWA Fig 221). At government request Order 795 of July 1917 built 26 new wagons of J18 type for use overseas. Given the diagram J19, they were fitted with Nord Railway screw couplings and safety chains (RWA Fig 225; cf number not cast on plate). Along with the CROCODILE J, they were the first bogie wagons to use large circular self-contained buffers rather than the oval-faced type. A photograph in 'The Great Tanks' by Ellis and Chamberlain (Hamlyn, 1975) shows Mark IV tanks loaded on these wagons before the battle of Cambrai, in November 1917 (see also 'Railways and War before 1918' by Bishop and Davies (Blandford, 1972) plate 197). Tanks at the end of the war were heavier than before, and 40-ton wagons were produced from the heavily trussed J18 design by the use of bogies with larger journals (J20); 84458 was altered from J18 in May 1917, but when a rush came for these wagons a year later there was no time to rebuild the bogies, so Order 1048 was issued 'to exchange the bogies of 30 MACAWS D [sic]

for those under 40-ton coal wagons and engineering department rail wagons' (Vol.1, pp. 22/56).

After World War I these vehicles were returned to normal service, J17 coming out of the diagram book in January 1919 when the bolsters etc were replaced on the 32 remaining military MACAWS B (Order 1275). The strengthened underframe wagons, which were fractionally wider than J17, became the new diagrams J22/3/4. It was originally intended that all of J20 should revert to 30 tons, but eight kept the 40-ton rating to become J22. The bolsters and side rails were replaced, but the floorboards were still exposed on these officially coded MACAWS D. The other 22 J20 wagons, plus the remaining 50 wagons of J18, became the 30-ton MACAWS E (J23). The overseas wagons J19 were repurchased from the government and were added to the MACAWS E (heavily trussed, but rated only at 30 tons). Order 1517 of August 1921 instructed 'to overhaul (fitting with bolsters, binding chains, and side rails) and rewrite 26 overseas MACAWS B [sic] WD 39201-26, with standard GW writing. To be marked MACAW E and to be renumbered GW 70719-44.' Evidently this was amended to give some MACAW E FLATS which ran without bolsters, having floors consisting merely of steel plate on battens (J24).

A 'light' 20-ton bolster wagon 35ft over headstocks was introduced in 1927 coded MACAW H with restored full curb rails (J25, RWA Fig 226). The wagons were smaller versions of the J21 MACAWS B introduced in 1917 for ordinary (not war) traffic, Vol.1, p.68. The principal difference in J21 from J14 was the large-headed self-contained buffing and drawgear. The MACAWS H were intended to replace the old MITES beginning to be withdrawn. The telegraph codes F and G had been taken by ex-TVR 10-ton four-wheel bolster wagons and 30-ton bogie bolster wagons respectively for which no diagram numbers were issued, both varieties of wagon being used without modification at Swindon; for example, the MACAW G retained diamond-frame bogies. About this time the remaining J2 BEAVER D 48902 which was comparable to the TVR wagons, was recoded MACAW G for simplicity; it had taken the bogies of its scrapped J3 partner 48901 and lasted itself until 1943.

As explained elsewhere the MACAW E FLATS and some MACAWS B were converted into BOCARS in the late 1920s and early 1930s.

Since the 70ft length of the MACAW C limited its use, especially on foreign companies' lines, a new rail wagon for both engineering and merchandise use was designed in 1935 with length limited to 60ft but uprated to 40 tons; 60ft rails had come into use in the late 1920s. The engineering department vehicles of this type took the new code GANE A but the ordinary stock became the new MACAW J (J26); curiously the 1939 *Telegraph Message Code Book* does not contain the GANE A code. For comparison, the CROCODILE M of 1938 was 62ft 6in over headstocks.

World War II repeated the same traffic problems as World War I and bogie bolster wagons and RECTANKS were sent to France. The acute shortage of bolster wagons led to BOCARS regaining their former identity as MACAWS E/B, and to the conversion of BOCARS A into the light MACAW Z (J27) in 1940 with bolsters, chains and stanchions. Incidentally, some of G28/9 had been reconverted before the publication of the 1939 *Special Wagon Book*, in which the list of MACAWS E is incomplete. At first, even the original

MACAW E FLATS were given bolsters and stanchions, but in January 1943 these fittings were removed from 20 vehicles of J23 to give FLATS as in 1917, thus re-introducing J24 (Vol.1, p.25). This facilitated the loading of large cases whilst still allowing the conveyance of Bren gun carriers and tracked armoured vehicles. The wagons were based at Swindon and were ordered through the Chief Goods Manager's Office.

Diagrams J28/9/30 were essentially J21/26/25 respectively, with two independent long lever brakes instead of DC after the decision in 1939 not to use the GWR brake on new construction (Vol.1, p.80). Finally J31 related to eight RECTANKS fitted with four bolsters and stanchions in 1947, coded BOGIE BOLSTER A (RWA Fig 100); of these postwar conversions, only 70750/2 had earlier been fitted with three bolsters before the War (see Vol.1, Chapter 3).

CHAPTER 9
K – CRANE TESTING WAGONS

These vehicles were engineering department tenders for the crane testing vans (for example CC4) that were sent about the system to proof test the various cranes and coal hoists. All vehicles had rings whereby they could be lifted off the track, cast-iron pigs providing the weights. Basically flat wagons, they had steel floors with curb rails, and were provided with special arrangements so that the wheels would not fall out when being lifted.

Four separate diagrams were issued, covering the period from 1892 to 1921. The first two wagons (subsequently K2) were equipped for running in passenger trains, with Armstrong vacuum system,

3ft 7½in wheels and long springs on an underframe similar to contemporary MICAS (X1) but with a 9ft wheelbase (RWA Fig 119 shows K2 in BR days). K1 built at the turn of the century were not vacuum-fitted, had 3ft 1½in wheels and were longer heavy 20-ton wagons (19ft over headstocks, 11ft wheelbase) with 1ft channel underframe; they were amongst the last vehicles to be fitted with the Thomas brake. The slings on these two wagons were attached to the inside underframe scantlings via two rectangular holes cut in the floor, whereas the rings were anchored on the solebars proper on all other wagons.

K3 (1909) and K4 (1921) were essentially the same design with DC III vacuum brakes and steam pipes, the underframe and buffing gear reflecting O14/18 respectively, except for long springs. Some drawings indicate that K3 was supposed to have a 'pyramid' frame mounted on the wagon, with a central sling; indeed, L612 talks of the truck as being 6ft high, but neither K3/4 were built with the feature. K4 was stationed at the London Divisional Engineering Department at West Ealing.

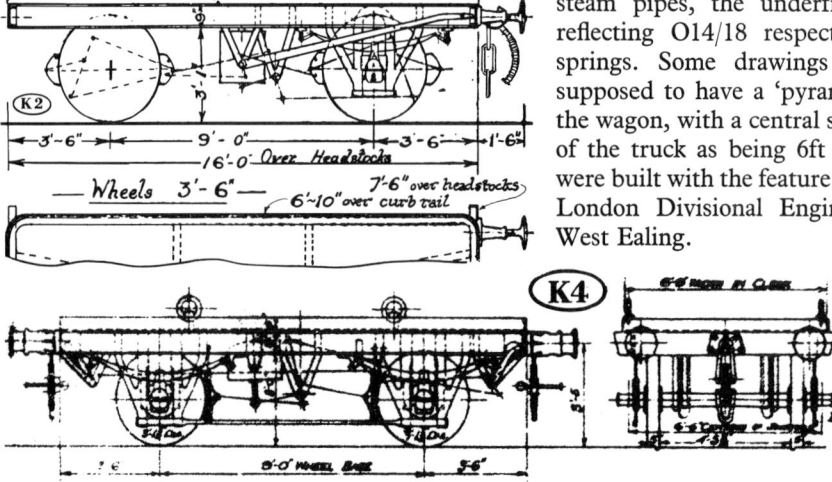

For K3, substitute GW laminated sprung buffers.

K3 built 1909, photographed 1965.

(W. Beard)

CHAPTER 10
L – MATCH TRUCKS AND WEIGHT TENDERS

This group was originally intended for crane match trucks only, since trucks which rode under long loads overhanging other wagons were typically J8 MACAWS or low-sided opens. When these latter types ceased to be common by the 1930s, flat match trucks were then converted from old wagons and incorporated in the L index for the first time (L21-3). The list was fairly stagnant since new cranes were introduced infrequently, and wear and tear of the trucks was not excessive. After the grouping, it became the practice to convert old

were five basic four-wheel types, see Table 1.

They were used for handcranes and steamcranes up to 12 tons, roughly with size in proportion to crane capacity. There was also a bogie type L11/17 from 1908/18 for the heavy 36-ton steam breakdown cranes, 37ft 6in over headstocks by 8ft 9in (9ft 2in over stanchions), with 5ft 6in bogies at 28ft centres. The superstructure of boxes along the middle of the wagon sides, with a central way for the jib, was common to all. Except on the 16ft over headstocks wagons, the boxes did not go the whole

TABLE 1

| Diagram Nos | Date | Over Headstocks | Width | | Wheelbase |
			Over Body	Over Stanchions	
L1/3/4	1896–1902	22ft 6in	7ft 6in	7ft 11in	13ft
L2/8/9/16/19	1898–1918	30ft 5in	7ft 6in	7ft 11in	19ft
L5/15	1903/13	19ft 6in	7ft 5in	7ft 10in	11ft
L6/7/13	1900-10	16ft 0in	7ft 6in	7ft 11in	9ft
L12/20	1908/26	33ft 5in	7ft 11in	8ft 4in	22ft

wagons into replacement crane match trucks rather than build new (cf stores vans). For example, lot 1101 in 1932 used the underframe of an old fruit and parcels van (1312) as the basis of a match truck for crane 130 at Barry. Likewise lot 1102 used covered goods 85915 for crane 546. Such rebuilds were not entered in the index, and ultimately all the crane match trucks were removed from the list, leaving only L21-3 as merchandise match trucks.

In the early days converted old tenders and 15ft 6in OPENS served as crane match trucks (such as L14), but eventually an iron pattern emerged which became standard with minor variations. Overhanging crane jibs rested on the trucks, so an axle arrangement on trunnions was usually provided, which allowed the jib seat to slide sideways when going round curves. The rest of the wagon was fitted out with tool, dunnage and appliance boxes.

Among the standard iron match trucks there

length, leaving end platforms for packing and such like (cf RWA Figs 303/9/10/15/6/26/7/8/9).

Also in the list were weight trucks, which were tenders for the bogie match trucks. The two built in 1909 to go with L11 (given L10) and that for the overseas heavy crane (given L18) were essentially modified J12 rail wagons in concept.

Lot 916 in 1923-4 converted a bogie match truck (purchased from the Port Talbot Graving Dock & Shipbuilding Co) into a four-bolster wagon, numbered 48997.

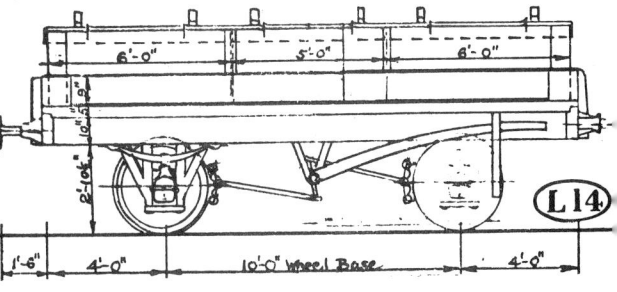

Above: *L18 built and photographed 1918. Weight truck tender for 35-ton steam crane for overseas. Similar to L10 except for self-contained buffers. Likewise L17 bogie match truck was s/c buffered version of L11. J-hangers, underslung springs, OK boxes, DC cross-cornered brakes (no white stripe). Flat trussing with adjustable kingposts. Fall down sides with removable stanchions.*

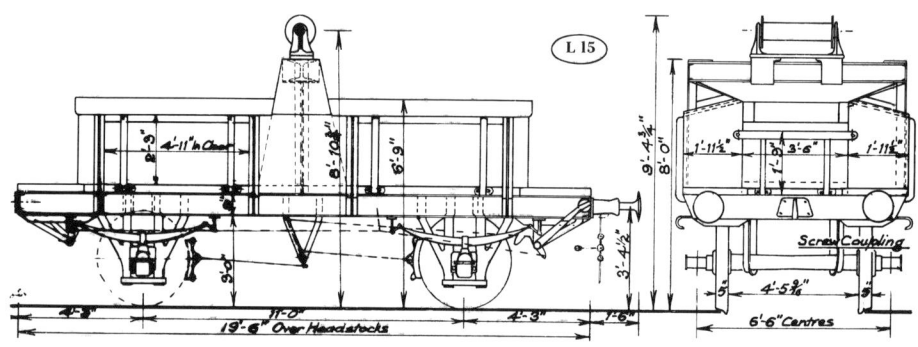

CHAPTER 11
M – SHUNTERS' TRUCKS

The GWR was the only British railway which constructed special shunting trucks in large quantities. After a few experimental designs, the wagon which was to become so well known first appeared in 1895, and the 14ft over headstocks, 7ft wheelbase scheme with tool/lamp box, deck handrails and stepboards was perpetuated with minor changes into nationalisation. Nearly 300 vehicles were built in all.

The early trucks had an independent lever brake on each side of the wagon and plate guards were provided in front of the wheels to prevent accidents to men riding the running boards. The advantages of the quick-action DC brake mechanism were soon realised, and L373 of 1902 was one of the vehicles fitted with the 'experimental' Mk I brake. It is in this form that M1 of the index is drawn. M2 dates from 1906 with the DC III brake

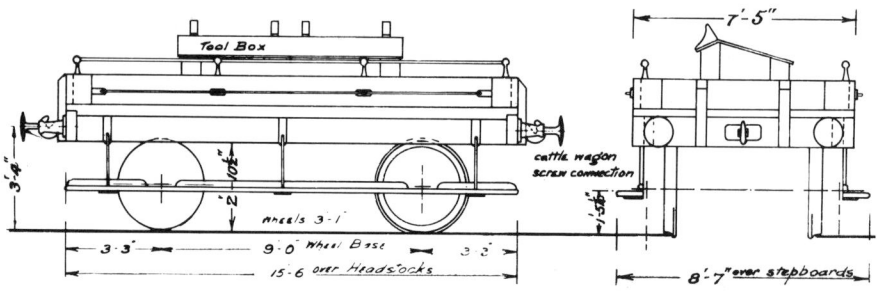

15ft 6in shunting truck converted from 19th century underframe.

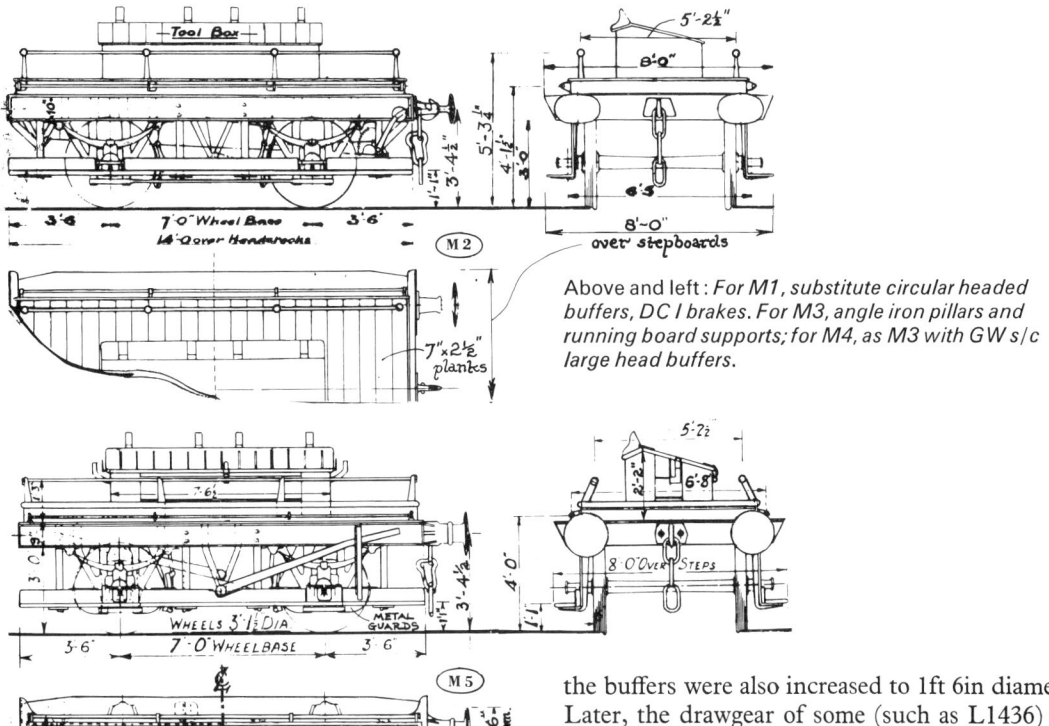

Above and left: For M1, substitute circular headed buffers, DC I brakes. For M3, angle iron pillars and running board supports; for M4, as M3 with GW s/c large head buffers.

and large oval-headed buffers. M1 had used ordinary wagon buffers, but the use of oversize buffers meant that wagons in all conditions could be accommodated. Additionally, the buffer centre line was set down to 3ft 4½in instead of 3ft 5½in; this followed on all subsequent shunters' trucks. Instanter couplings were used after 1908. Many of of M1 were converted to DC III and large self-contained buffers in later years (RWA Fig 191).

The channel underframe was reduced from 10in to 9in on M3 in 1912, bringing the handrail height down further to 5ft 2in above the track. The handrails around the deck of the wagon had, up to this time, been fashioned from elegant pillars and round rod, but on M3 angle iron was substituted for the pillars. Self-contained buffers in the large 1ft 4in diameter heads were introduced in 1917 (cf C8/J19). This was the basis of the M4 diagram; the modification from M3 was trivial as seen from the relevant drawing numbers 48524 and 48524ᴬ. Subsequently the M3 diagram was discontinued and all those trucks lumped in with M4.

After 1939 the underframe had to be altered, along with all other designs, to replace the DC brake. Thus M5 had the familiar M4 body, but with the Morton either-side brake; the heads of

the buffers were also increased to 1ft 6in diameter. Later, the drawgear of some (such as L1436) was of the short variety, as on OPENS in the middle of the war.

The journals on all the M group were an odd small size, 8ft by 3⅝in. L180 of 1897 was the first truck to be fitted with these oil boxes, but it became the practice not to fit white metal linings to shunters' trucks in view of the 'stop-start' nature of their work.

Shunters' trucks were usually built for a specific place and had the depot name painted on; the fitting of vacuum and steam pipes depended on the location to which the truck was being sent and on what sort of work it would do. Steam and vacuum fittings were needed for carriage shed shunting, and vacuum alone for express freight train shunting. Some allocations were as follows: Acton M4 41048; Acton East M1 41811; Avonmouth Docks M1 41771, M4 94976; Bassaleg M5 41845; Bilston M2 41752; Birmingham Moor St M4 94947; Bordesley M4 41051; Brentford M4 41047, 94949; Bristol M3 94983, M4 41052; Briton Ferry M1 41809; Cardiff M1 41898, M4 41050/4, M5 41157-60; Cheltenham M3 41735; Didcot Ordnance Depot M5 41087/8; Didcot Provender Store M1 41853; Fishguard M5 41817; Hockley Goods M5 41089/90; Laira M1 41883, M4 41046, M5 41883; Llandilo M4 41049; Llanelly M1 41810; Newport, Alexandra Dock Junction M1 41770; Newton Abbot M3 41734; Neyland M2 41739; Old Oak Common M4 41053; Oxley M1 41833; Paddington

M1 built and photographed 1894. Earliest standard type with grease boxes and single-sided lever brake. Laminated sprung buffing and drawgear. 10in solebar. White destination card frame on label box. Cast plates for number and GWR, but painted tare necessary since no load plate. Rectangular plate guards in front of wheels. Pillar handrail supports. Round stepboard hangers.

M4 built and photographed 1937. GWR oil boxes, disc wheels. DC cross-cornered brakes (diagonal central white stripe). Vacuum pipe. Large circular self-contained buffers. 9in solebar. Label clip.'Oval' shape works number plate with merely 'G.W.Ry' and number. Plate guards shaped to wheel outline. Angle iron handrail supports. Flat stepboard supports.

M1 41822; Park Royal M4 41055; Pontypool Rd M3 41738; Rogerstone M3 95000; Swindon M1 43957, M2 43904 (Loco Works), M4 43960 (No. 3 Carriage Shop), M3 43906, M4 43975 (Wagon Works); West London Carriage Sidings M1 41828-30; Weymouth M1 41831.

It is remarkable that the last shunting trucks made by the GWR were replacements for some of the earliest M1 designs that were still faithfully performing their duties fifty years on, often in their original locations. For example, 41817 had been sent in 1899 to 'New Milford' and a new wagon with the same number was sent to Fishguard in 1946. Likewise 41898 went to Cardiff in 1895 and 1946; 41829 to West London Carriage Sidings in 1899 and to Old Oak Common in 1946; 41823 to Worcester in 1899 and 1946. The first oil axlebox shunter's truck 41845 had been sent to

Birkenhead in August 1897; the replacement went to Bassaleg in 1946. Again 41834 for the Birkenhead Extension Lines in 1897 (RWA Fig 188) which was 'not suitable and returned' was replaced at Hollinswood in 1946.

Some shunting trucks were converted from old two-plank open wagons at the turn of the century. These 15ft 6in over headstocks, 9ft wheelbase wagons were not given a diagram number. How many were adapted is not known, but M4 L880 94949 for Brentford was built in 1922 to replace a shunting truck converted from osL53 21001 originally built at Saltney in 1872 (cf RWA Fig 187). It became the practice to convert old underframes into shunters' trucks after nationalisation, and the truck preserved by the GWS at Didcot is an old ex-V14 9ft wheelbase MINK A built in 1923.

CHAPTER 12
N – COAL AND MINERAL WAGONS

Apart from parts of the Central and Northern areas of the GWR, which received coal from pits in the Midlands and North Wales, most locomotive coal used on the GWR was steam coal from South Wales. The excellence of this coal is well known, and it was not uncommon at the turn of the century to see colliers in the Cardiff docks loading shipments for the Paris, Lyons et Méditerranée and other foreign railways.

There was a general scheme of regular distribution, a certain tonnage being allocated to each locomotive depot or engine shed. Large depots like Old Oak Common and Bristol received about 3,000 tons weekly whereas at the other end of the scale

coal was transferred direct from the wagon to engine in such places as St Ives, where about 20 tons per week were required – or Brixham with only 1 ton weekly. In all, there were about 140 coaling stations on the Company's system.

Special scheduled locomotive coal trains ran daily, one from Rogerstone to Old Oak Common, one to Bristol and another to the West of England, other wagons being attached to ordinary goods trains; similarly there were special locomotive empty wagon trains. As the wagons were emptied at the stages they were labelled to marshalling yards (in easy reach of the collieries), the principal of which were Rogerstone and Severn Tunnel unction. The collieries also advised each morning he Company's Central Coal Office at Pontypool Road how many loco empty wagons they had on hand and how many more they required for the day's loading. In turn the collieries were informed, by telephone, the sheds and depots to which the loaded wagons were to be labelled. The number of wagons to be sent to each particular shed was based on the agreed schedule of distribution, the tonnage being varied to meet the running of special trains and alterations in booked services. An emergency stockpile of about 30,000 tons was kept at Swindon; it began to be used in January 1944 because of the war.

Most merchandise coal was carried in gaily inscribed private owner wagons, a legacy of the days when canal owners did not provide barges themselves, but merely levied tolls for the use of their waterways. For transporting loco coal, however, the railway, although it did not own any mines, had its own wagons designed for use in the company's coaling stages. Many of the PO wagons were of small capacity, which led to inefficiencies in hauling. Thus there was a great drive, instigated by Sir Felix Pole in the 1920s, to encourage pit owners to use 20-ton mineral wagons, and by World War II over 5,000 high-capacity wagons had been built for hire to collieries. These wagons were grouped into the N index.

10- and 12-ton wagons

The oldest coal wagons in the diagram index were five-plank wooden wagons dating from the 1880s

(N7). Basically they were the then standard four-plank open with the addition of a wide extra plank to give 3ft 0⅜in inside height (RWA Fig 79 in 1904 painting trials). More than 500 of these wagons were built, but since they were soon superseded by iron wagons, L428 in 1904 related to 450 being fitted with sheet supporters and put into ordinary goods traffic (no O diagram was issued however); the rest were fitted with coke rails. Only seven remained in service in 1939, but some survived into nationalisation. The oldest iron coal wagons in the list were N6. Some of them (L47 in 1894) were built as ballast water tanks in DD2, by the addition of a tank top. N6 had a wooden floor, something not seen again on GWR iron coal wagons; there was a similarity with P4/5. The original 2ft 6in inside height of the 8- 9- or 10-ton version as built (RWA Fig 269), was later increased to 3ft 3in by the addition of a 9in plate rail round the top to give 10 tons, and this is the form illustrated in the wagon diagram index; RWA Fig 80 shows the large lettering as introduced with serifs on the G. Constructional features of these iron wagons related to round rolled corners, and joints in the $\frac{3}{16}$in side plates at the doors and behind the end T-stanchions. There were no exterior stanchions on the side of the wagon. Various other wooden and iron 'loco wagons' had been produced before N6 and N7, but their dimensions listed in the lots do not agree exactly with N6/7.

Replacements for N7 were N13 of 1905. Essentially these were modernised N6, with 'complete' sides 3ft 1½in in clear, DC I brakes and external T-stanchions at the doors. They were 8ft wide

N30 introduced 1935, photographed 1937. GWR 8 by 3¾in oil boxes, but RCH small rib buffers and other details. Morton brake. Eight-spoke wheels, one of the last vehicles painted in 16in letters but note that the label clip has already been repositioned over the left axle. Thinner numbers and 'Loco' than before.

N4 built and photographed 1900. First 20-ton iron loco coal wagon design. 1ft channel solebar. Thomas brake (horizontal OFF/ON plate). Four brakeblocks (early lots only two) with rigid connection at shoes. No external side stanchions. Turton's self-contained buffers, non-Gedge hook. Ten-spoke wheels. Legend on plates 'NOT TO BE LOADED WITH ASHES' 'GWR LOCO DEPT' etc.

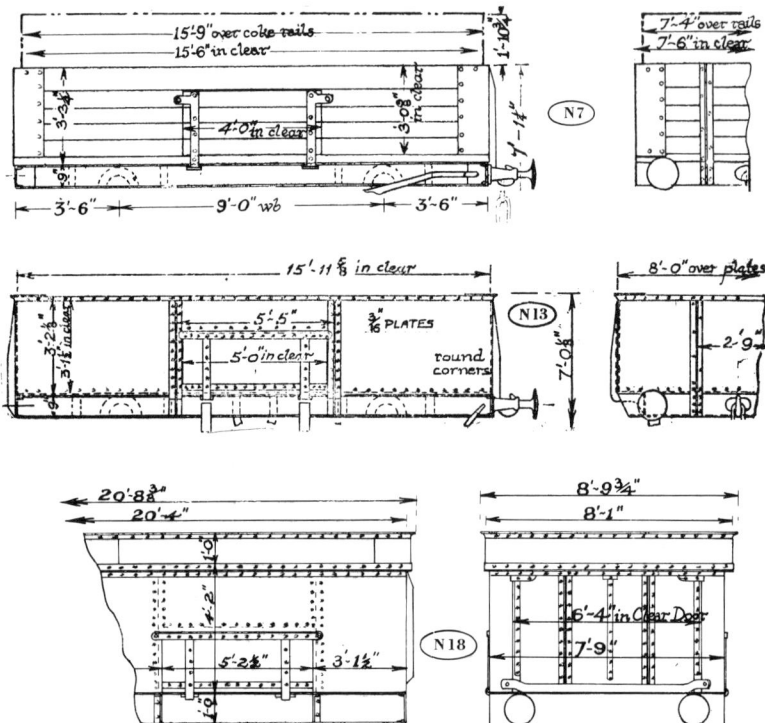

Top: *Underframe as on V6/P13.*

Centre: *Underframe as O4. For N6 as built, lower sides to 2ft 9in, narrow width to 7ft 6in, omit side T-stanchions (as N4), substitute wooden floor, substitute side lever brake u/f as on P13/V6; later add 9in plate rail round top. For N16, as N13 with sides increased to 3ft 9in in clear (3ft 9½in over coping). For N19, body as N13 (with 5ft 4in wide door), substitute DC III u/f with GW s/c buffers as on O15 (but non-vac). For N20, as N19 but square corners. For N21, as N16 (with 5ft 4in wide door), and square corners. For N30, like N20 but 6in longer, RCH fittings.*

Bottom: *For N3, increase height from 4ft 2in to 4ft 6in, modify capping as on N22. For N18, add top rail and end tip door as shown.*

rather than 7ft 6in; and iron floor plates ¼in thick were employed. These round-cornered wagons were produced until 1913, variations occurring with N19 where DC III brakes appeared, and the last 120 vehicles of this diagram had self-contained buffers also. N19 (for example 9950) had a slightly wider door, 5ft 4in in clear instead of 5ft. Soon after the introduction of N13, one wagon of L551 was built in 1906 with 3ft 9in sides rated at 12 tons; it was represented by diagram N16 and had 8in by 4in journals rather than 8in by 3¾in.

The 'round corner' feature disappeared in World War I and N20 issued in 1915-16 was for a square-cornered version of the 10-ton N19 (RWA Fig 74). Likewise N21 from 1918 was for 12-ton wagons, essentially N16 with square corners and the wider 5ft 4in doors (RWA Fig 73). The underframes of N20/1 were similar to N19, except that the V-hangers were offset. This might be because N20 were constructed from condemned vehicles. N21 L847 were the first vehicles to be finished with the 'new pattern' (self-contained) drawgear (cf military MACAWS B).

The final design of GWR 12-ton loco coal wagon was N30 of 1935, and 200 of these had been built by 1937. The body was that of N20, with slightly different corner capping, lengthened by 6in to give 16ft 6in over headstocks; the underframe, however, retained a 9ft wheelbase. RCH buffing

and drawgear were used, and in particular the Morton brake with toothed rack appeared. The same underframe design was subsequently employed on 8cu yd ballast wagons P18/20 after 1941. The side stanchions of N30 came straight down, not being tucked under the solebar as was the current fashion. All these small wagons might be used to supply branch-line loco depots; the coal capacity of a 5700 class pannier and the like was about 3½ tons.

An 'odd' diagram added to the index after the 1939 DC brake appraisal was N33, which was a group of more than 100 wooden six-plank 12-ton coal wagons. To RCH South Wales end-tipping private-owner specifications, they had been purchased from the Bute Works Supply Co in 1907 and were not typical of GWR design for metal or wooden coal wagons; the LMS and LNER, it may be pointed out, archaically adopted a wooden bodied, wooden solebar design for their standard mineral wagon.

Although special coal wagons were not normally reserved for GWR steamships, L1150 of 1934 related to the conversion of O24 OPENS for the Waterford steamers and L1262 of 1937 built coal container wagons for the SS *Great Western* at Fishguard (cf 10 ton grease box wagon in 16in lettering, RWA Fig 14).

This pattern of iron wagon development –

above: *N4 part of 36-wagon train with 'new double rake' (DC III conversions), photographed on dynamometer van trials, 7 February 1902 (van built on 293 1900–1, 8-wheel, no.790). Hence N4 wagons, a couple of years old equipped with vacuum brakes. Note BG style lettering, reintroduced after cast plate period, but before 25½in large GW in 1904 (cf N1/11 pattern wagon, vol.1, p.66).*

Below: *N24 built 1925 with Morton brake, converted in 1926 to DC brake, photographed 1936. No end number, 'GW' overlaps capping rail.*

round corners/no stanchions, round corners/ stanchions, square corners/stanchions, plus changes in underframe ideas – was also followed in 20-ton wagons.

20-ton wagons

The first GW 20-ton iron loco coal wagons were 20ft over headstocks by 7ft 9in by 4ft 2in high with two doors on each side (N4). Mounted on massive 8ft channels, the round-cornered body was constructed in a manner similar to N6. A 12ft wheelbase was used, and the brake fitted originally in 1899-1901 was the Thomas type, later supplanted by DC II and III brakes (cf RWA Fig 81, DCII and Fig 78, DCIII in 1904 painting style change: note serifs on G, and differing layouts). In 1901 a modified version of these vehicles appeared (N3) with the body height increased to 4ft 6in but the underframe channel reduced to 10in. The height above railhead of the coping (which was 'flared' instead of box-like) thus became 8ft 4in instead of

8ft 2in. One vehicle of the new diagram thus issued was later built with a still higher 4ft 10in body and external T-stanchions in March 1903, at about the same time that the bogie loco coal wagons (qv) appeared with stanchions. To maintain 5ft 2½in clear at the doors, the panels became 5ft 7½in between centres of T-irons. In fact the centre panel was widened to the same size, so that the four side stanchions were symmetrically placed with 1ft 6¾in spacing from the ends of the wagon. The Swindon diagram (N5)* shows an additional door in the centre panel (giving three doors per side) and it is not clear whether this feature was a later addition (cf N28/9). Probably this vehicle was the pattern vehicle for stanchioned 20-ton coal wagons, but a decision was made in the interim to lengthen the underframe to 21ft over headstocks, since it

*The running number of N5 is not known. However, it was *not* on L525, as sometimes suggested, which related to the purchase of a 21ft 6in oh, 9-plank *wooden* loco coal wagon from the Gloucester RCW, which was not given a diagram number.

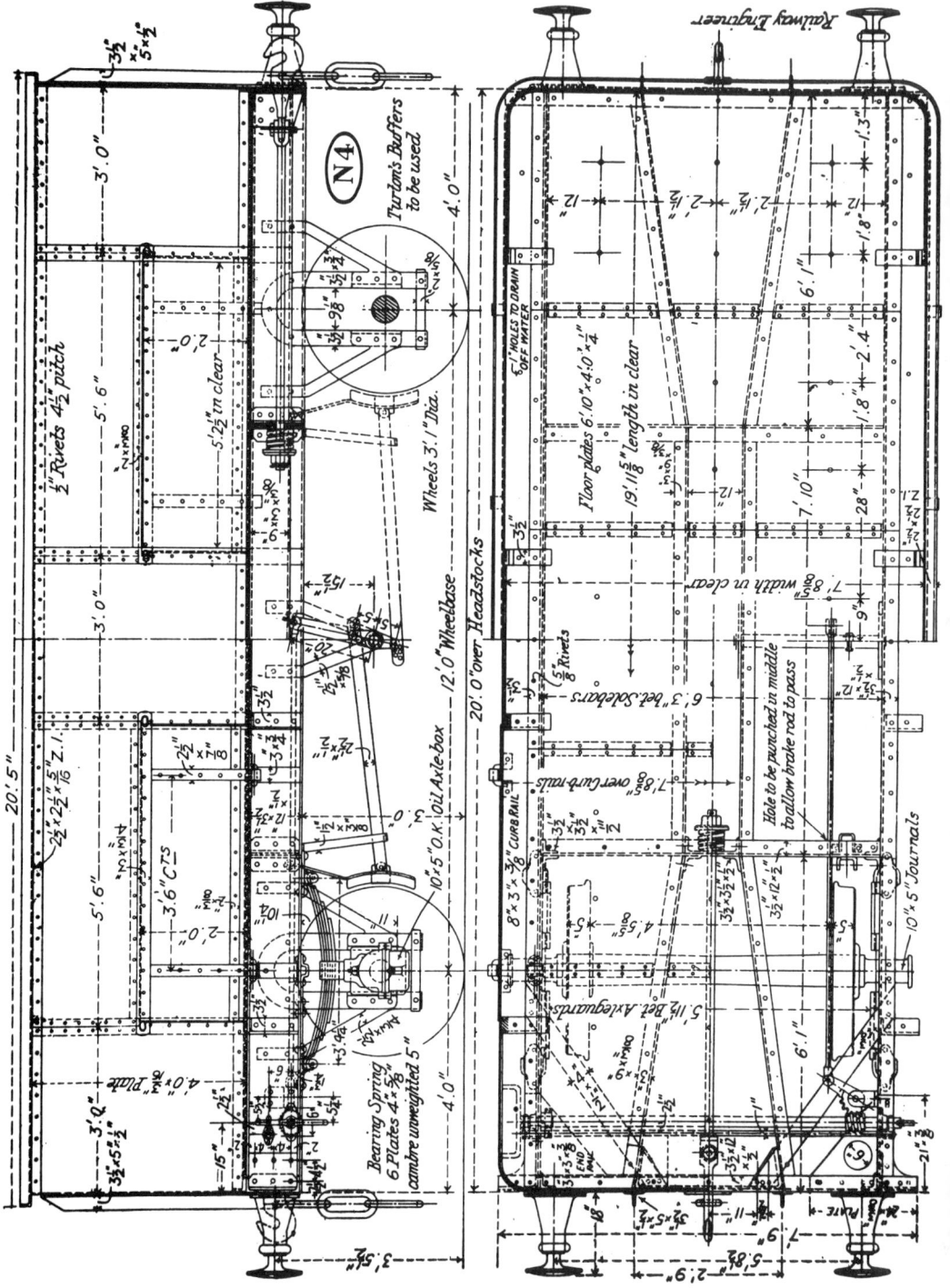

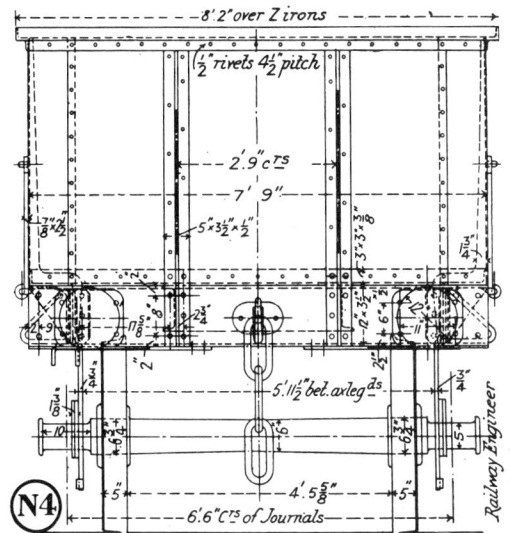

N4 20ton iron loco coal wagon. Drawing to 8mm scale.

was a 21ft stanchioned design that was produced in large quantities early in Churchward's time. The diagram was N2, and the body reverted to N3 proportions, the 1ft stretch appearing in the centre panel between the two doors (Vol.1, p.74; RWA Fig 66). N2 wagons were built at regular intervals until 1913 and a further 150 vehicles appeared in 1923-4; the latter had a modified underframe so were given the index N22. The first lots of N2 had DC II brakes, like N3/5, but later they were equipped with the Mk III brake, which was the version shown in the index, the suffix on the drawing number (24931[A]) indicating this modification. Express freight trains were in vogue at this period, and some of N2 were vacuum-fitted (as were N4 when their Thomas brakes were replaced) but the practice was not continued, and the fittings were removed in 1934. N22 was curious because smaller capacity square-cornered wagons had been introduced some years earlier and proposed alterations to square corners and top capping angles for future 20-ton coal wagons had been made in 1918; the underframe changes on N22 related to V-rather than crowned axleguards, and to GWR self-contained rather than laminated-sprung buffers.

In the N-group were classified mineral wagons as well as loco coal wagons; in this case 'mineral' principally meant private-owner industrial coal and coke rather than stone chippings, iron ore, or the like.

Much had been written from the turn of the century on the economics of large wagons.

Although success may have been limited with large merchandise vehicles, it was foolish for so much of the 50 million tons annual production of Welsh coal to be carried in 9-, 10- and 12-ton wagons, which required special shunting. Large wagons also saved siding space and maintenance costs. The GWR management made a strong effort in the summer of 1923 to persuade colliery proprietors to use 20-ton wagons; £285,000 was spent on 950 new end-tip wagons, and over £2 million on alterations to 160 or so quayside coal shipping installations. Special rebates on conveyance rates were instituted and dock charges substantially reduced if the new wagons were employed; firms could also buy the wagons on hire-purchase. Mr J. Auld, Docks and Personal Assistant to the Chief Mechanical Engineer, in a paper read before the Institute of Transport in January 1926, explained that the 20-ton wagon was chosen because it was the largest capacity mineral wagon that could usefully be carried on four wheels, given the condition of colliery sidings. Table 2 demonstrates that for a 600-ton train, savings in terms of siding accommodation and dead weight using 40-ton wagons rather than 20-ton wagons were only marginal, and the capital cost per ton of load of bogie wagons was higher. Inciden-

TABLE 2

Wagon Capacity Tons	Number of Wagons for 600 Tons	Length of Train Feet	Tare Weight Tons
8	75	1350	400
10	60	1080	369
12	50	975	350
20	30	735	288
40	15	690	280

tally, although the Standing Committee on Mineral Transport (set up following the 1926 Royal Commission on the Coal Industry) advocated the standardisation of the 20-ton wagon, the 10,000 steel mineral wagons built in World War II for general use were 16-ton capacity.

The GWR mineral wagons were square-cornered iron wagons, 21ft 6in over headstocks (6in longer than N22), but they retained a 12ft wheelbase to accommodate the tight curves of colliery sidings; they also were wider than before, 8ft 4in over capping and 8ft 6in over stanchions and doorstops. Although built to RCH specifications, the 1907

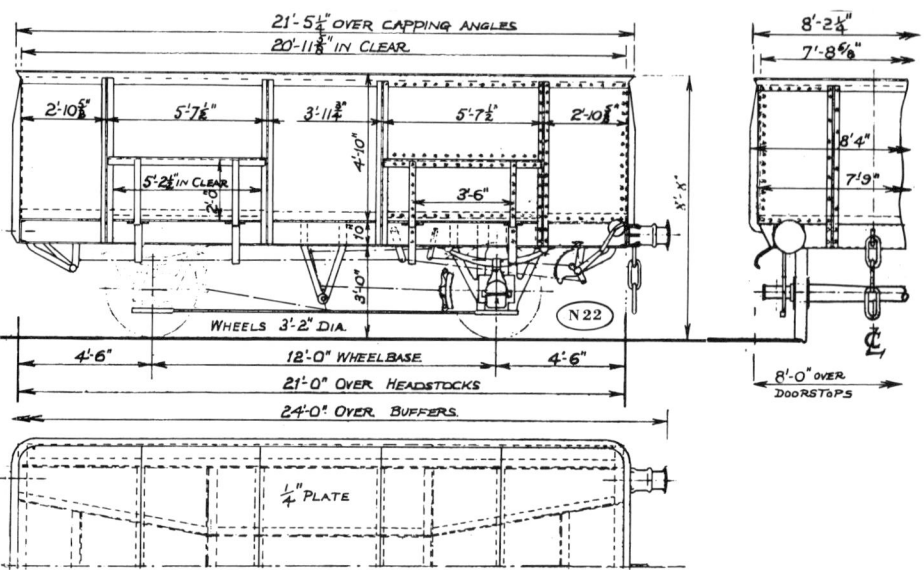

For N2, substitute GW laminated sprung buffers, and crowned axleguards.

Below:
For N27, body of N24 with 'solid' ends and T-stanchions, u/f of N29, but GW s/c buffers. For N32, side view of N24, one solid end of N27, one new reinforced tipping end (as shown), slotted link Morton brake as V20/P19, RCH buffers. For N34, as N27, but RCH buffers and modified Morton as N32 (also built as 21 ton).

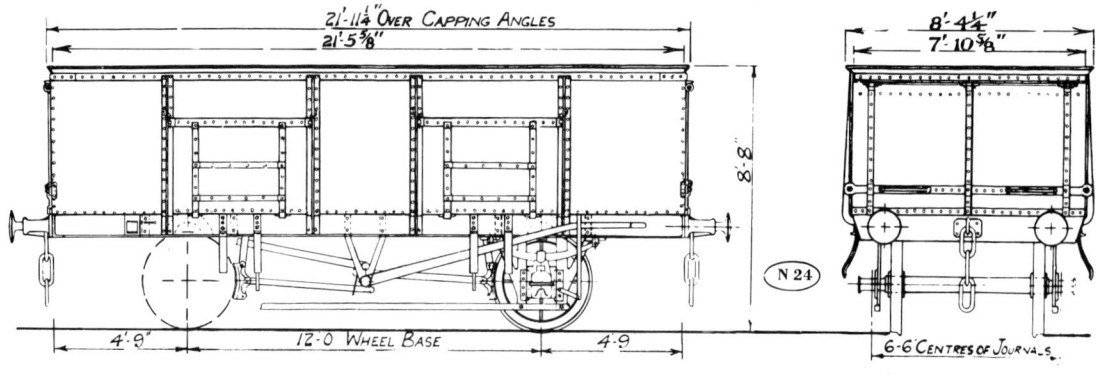

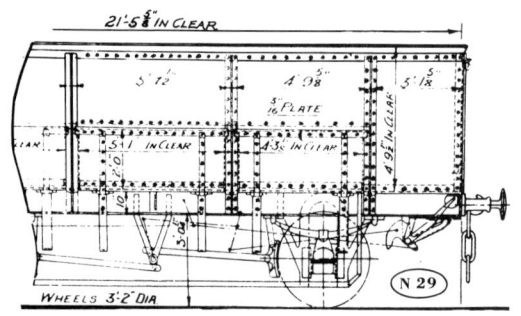

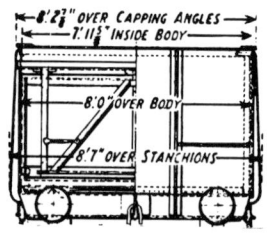

N32 end.

End as N24. For N28, body as N29 but revert to original Morton u/f, as on N24. For N23, as N28 omitting extra doors, ie only central side door. For N31, body of N24 u/f of N29.

regulations (BGW Appendix 5) state that 'no wagon for tipping coal in the Bristol Channel ports must be of a greater width than 8ft 3in'; the new width presumably reflected the 'erection of new coal hoists'. The tipping feature for shipment traffic was new on GWR-designed wagons; the axles were strong enough to tip to 45° but 50° was considered an abuse. As a prelude to the introduction of N23/4, two wagons from N4 (53190 and 53312) had been converted to end tipping in 1909. An additional 1ft rail was eventually put on the sides giving a 9ft 2in height above rail level, and a door was provided at one end; because of the round corners the door was only 6ft 4in in clear. The wagons (N18) ran between Glenavon, Llynvi and Swansea East Dock and were intended to test the Tanner-Walker hoist, used for loading coal on to ships at Swansea.

The Pole wagons were not introduced without market research, and many were set aside for specific firms such as North's Navigation Collieries Ltd, D. R. Llewellyn, Merrett & Price; Bedwas Navigation Colliery Co Ltd, Crumlin Valley Collieries Ltd; W. D. Rees & Co Ltd; Stockwood Rees & Co Ltd. The first 500 were used on shipment coal at Port Talbot and King's Dock, Swansea, and later ones were used at Newport Docks. Although these wagons were intended for industrial use, some private coal merchants had them. Depending on the needs of the firms, two versions of the wagon were produced: 760 vehicles with one door per side (N23 Vol.1, p.78) and 211 with two doors per side (N24/31), both with tipping doors at each end. They were built under six lots, the 200 wagons of L931 being constructed at Swindon, the rest (L936-8/84/1040) by Gloucester RCW, Birmingham RCW, Bute Works Supply Co/Stableford & Co, Cambrian Wagon Co, and Fairfield Shipbuilding & Engineering Co, respectively. RCH details prevailed, N23/4 being amongst the first GWR wagons to be fitted with RCH coil spring buffers and through drawgear; the ribbing on the buffers varied with the different outside builders. Morton brakes were used on all of N23/4 originally, but on Order F452 of January 1926, DC III brakes were substituted on 100 of the Swindon-built N23 in order to compare the per-

formance of the two types; L984 purchased one DC III braked mineral wagon and 20 more were obtained in 1929 on L1040. The Swindon conversions (with one door per side) were not given a new N index number, but those built by contractors (with two doors per side) were later given N31.

It was logical that new 20-ton loco coal wagons would take the Pole design, and N27 of 1925 was essentially the body of N24 with 'solid' ends; the underframe, however, used GWR self-contained buffers and DC III brakes (vol.1, p.52). The design was produced until 1938, and foreran N31 regarding brakes. New construction after 1939 had to use Morton brakes, so N34 from 1944 into nationalisation was basically N27 with a modified Morton arrangement (slotted link instead of cam clutch) and RCH buffers (cf N32). These last loco coal wagons were built at 21 tons as such; all 20-ton wagons younger than N23 were uprated in World War II. N27 wagons were built fairly regularly between 1925 and 1933, but then came a gap until the final lot in 1938. After N27 L1103 in 1933, L1127-36 inclusive related to new mineral wagons (N32), and the 5,000 vehicles constructed by various outside builders were let out to the collieries on 'redemption hire' (hire-purchase). They did not bear GWR running numbers, receiving H1001-H6000, and were not in the GWR stock books. The body was N24 with modified capping, but with only one tipping end, the door of which had diagonal stiffeners additional to the pattern of the other end-tip wagons; stiffeners had indeed been added to some of N23/4 after 1928, as some bulging of end doors had taken place. The doors were closed by lugs rather than drop pins. The underframe of N32 was RCH with modified Morton brake, the arrangement which N34 took in 1944. The £1 million cost of N32 came from the interest accrued on the postponement of certain work on the GWR system that had been authorised under

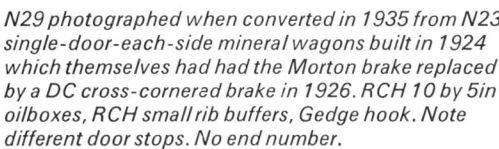

N29 photographed when converted in 1935 from N23 single-door-each-side mineral wagons built in 1924 which themselves had had the Morton brake replaced by a DC cross-cornered brake in 1926. RCH 10 by 5in oilboxes, RCH small rib buffers, Gedge hook. Note different door stops. No end number.

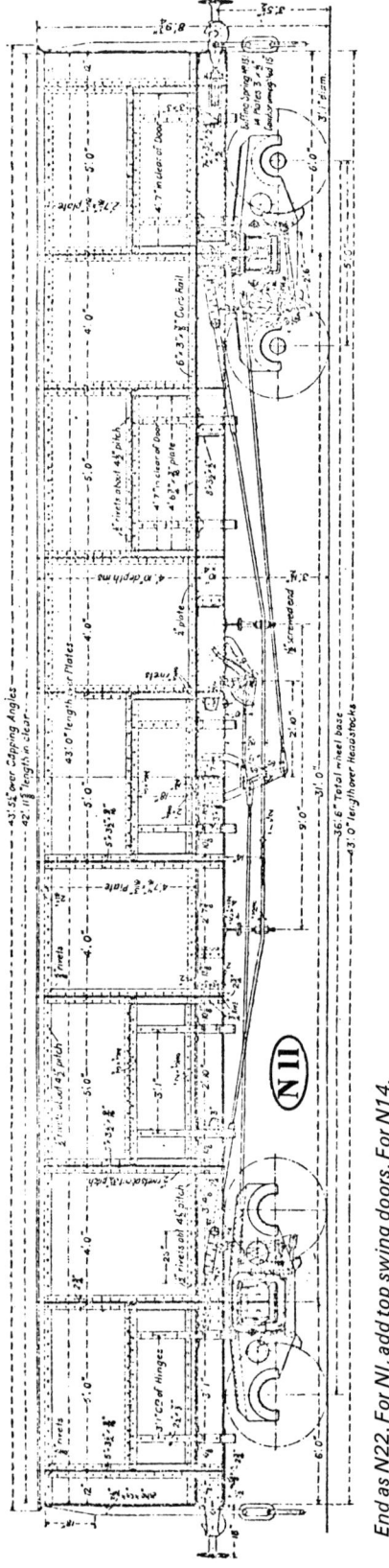

End as N22. For N1, add top swing doors. For N14, widen to 8ft 4½in (7ft 10⅝in in clear, 8ft 6in over all), substitute two separate DC systems each end. For N15, as N14 but wheelbase lengthened to 28ft from 25ft 6in, DC levers inside wb. For N17, body and u/f as N15 but 'central' DC brake as on N11. Not to 4 mm scale.

the Development (Loan Guarantees and Grants) Act of 1929. With the six-fold multiplication of the mineral wagon fleet, opportunity was taken to convert 200 N23 wagons to loco coal wagons by the provision of additional doors, 4ft 3in in clear, in the panels adjacent to the original door, making three doors per side. This was done under two lots, 1154 and 1217, in 1934-5 and explains the gap in the building of N27 wagons. Some of the converted vehicles were those that had had the Morton brake exchanged for the DC III brake in 1926, so two diagrams were issued, N28 for 149 Morton and N29 for 51 cross-cornered; both types were altered under each lot, and some renumbering took place.

Bogie coal wagons

The 40-ton bogie loco coal wagons introduced by the GWR in 1904 were among the largest loco coal wagons built on any British railway. Since the GWR was both customer and consignee for loco coal traffic and was not bound completely by clearance and weight restrictions, it need buy only from mines where its wagons were acceptable. Long before the efforts of Sir Felix Pole to use larger wagons for traffic coal, the GWR had 'practised what it preached' in using these even larger wagons for its own supplies.

At the time of their introduction the *Railway Engineer* commented as follows:

> The tare of these 40-ton wagons, which are constructed of steel throughout, is 18 tons 13 cwt., while that of a 10 ton coal wagon averages about 6 tons 1 cwt., so that on the haulage of dead weight there is a considerable economy. These wagons will also form a useful object lesson to private owners and serve to familiarise colliery owners, shunters and others with their use. As we have often shown the private owners have no direct financial interest in either reducing the tare of wagons or increasing their carrying capacity, and they are able to urge cogent reasons for not doing so, one of the chief of which is that it does not suit their customers to take their coal in large lots any more than it suits a railway company to take its locomotive coal in 40 ton wagons. For some few years some railway companies have been making every endeavour to spoil this line of argument by using wagons of large carrying capacity, and these large wagons that Mr. Churchward has built are another step in that direction.

However, by the grouping, policy had swung away somewhat from the large bogie wagon, as suggested by Mr Auld's paper.

Twenty-seven wagons were constructed in the period 1904-10, under four lots, for which five diagrams were issued (N1/11/14/15/17). The bodies were designed in conjuction with N2 as seen by the sequential drawing numbers; as such, the first lots had 4ft 10in high side plates by 7ft 9in width (N1/11, vol.1, p.66). On later lots however the width was increased to 7ft 11in over plates. All the wagons were 43ft over headstocks and had five discharging doors on each side 4ft 7in by 2ft in clear, the panels containing doors being 5ft between T-stanchions, and the remainder 4ft between stanchions. Various underframe DC brake layouts were tried out on these bogie loco coal wagons, for it had yet to be established on long vehicles whether independent DC brakes on each bogie or a linked system was desirable. (Vol.1, pp.76/7). The brake layouts gave rise to different distances between bogie centres, and these differences, among other things, account for the five diagrams. The pattern wagon 54000 was delivered on L420 in March 1904 at a cost of £338.15.2d, having been ordered in February 1903: it had brakes on both bogies linked to a single cross-shaft and DC ratchet mechanism under the middle of the wagon. The distance between bogie centres was 31ft, so that the axles of the outer wheels were 3ft 3in in from the headstocks. In contrast, the first V1 MINKS F of 1902, and the J11 MACAWS B of 1904 on L450, had only one bogie DC braked as originally built, with the levers at the headstock, similar to bogie CROCODILES. Also on V1 the bogies were closer to the headstocks than the first bogie loco coal wagon (2ft 9in instead of the 3ft 3in); on J11 they were closer still, at 2ft 6in. Ten more bogie loco coal wagons identical to 54000 as built were ordered on L507 in

July 1905. However, an alteration was made to 54000 in November 1905 to incorporate top swing doors above each side door, a suffix A on the print number 22603 indicating the change. Lot 507 was still in construction at that time and six of those vehicles eventually received top doors like the rebuilt 54000; L507 was completed in June 1906. The N group of the diagram index was evidently drawn up in the middle of this period, for the top-door version was placed as N1 and the other as N11. Another print number 24932, with a suffix A for the top-door emendation, is given in the Swindon diagram book for the N1 versions of the L507 N11 vehicles.

No further versions of the bogie loco coal wagons had top swing doors, but three more diagram numbers were issued (N14/15/17) to cover brake, bogie centre distance, and width changes. Whereas N1/11, like N2, were 7ft 8⅝in width in clear (8ft 2¼in over capping), all the rest were 8ft 2in in clear (with the corresponding over-capping distance increased to 8ft 4¼in). On N14 the brake arrangement consisted of two separate DC systems on each bogie, with the ratchet, cross rods and levers at each headstock. On N15, the bogies were moved further apart, the outer axles being 2ft away from the headstocks, with a bogie centre distance of 33ft 6in. The contemporary MACAWS B of L553 et seq (J4) had also moved the bogies closer to the ends of the wagon, giving a 2ft distance. There was no room for the two DC cross-shaft gears at the ends of these wagons (unlike N14), so on N15 the two separate brake levers were placed inboard of the

N1 built 1907, photographed 1935 with 16in letters. Top swing doors, necessitating 'Loco' to be moved to asymmetrical panel on right. Central DC II system for both bogies. Square shank oval buffers. Ten-spoke wheels. Wagon renumbered.

N2/14 both built 1905. Photographed 1910 loaded with coke. N2 DC III brakes by this time (central white stripe), but was from L481 built without vacuum brakes. T-stanchions, laminated buffing and drawgear. 10 by 5in OK boxes. White patch on label box. Ten-spoke wheels. N14 (53993) had two separate DC II systems at each headstock as part of trials for bogie brake layouts. Four rows flat trussing with adjustable kingposts. No external stiffeners on bogie, 10 by 5in journals. Round guide oval buffers. Someone has chalked the script GWR monogram on the left hand door of N2!

bogies. It is worth observing that the MACAWS B on the following L553 had the 'conventional' DC system, earlier single bogie brakes on J11 being replaced by the central system. The last lot of bogie loco coal wagons (L643, 53980-89) were built in 1910 with the wider body of N14/15, and with the wider spaced bogies of N15, but with only one central DC system linked to both bogies (as on the original N1/11 scheme, the difference from which of course related to the bogie centre line distances).

Apart from 54000 itself, and the ten vehicles of N17, no mention has yet been made of which vehicles from which lots were in the N1/11/14/15 diagrams. The vehicles in question come from L507 (53990-99) and L552 (30495-500) ordered January 1907 and completed by July 1907. As given in Vol.1, Chapter 3, it was believed originally that some of L552 were the vehicles (in addition to L420) built or rebuilt with top swing doors, and additionally renumbered into the 539xx series. In particular it was believed that 30500 was in N15, and the rest of L552 in N1 with top doors. The reason for believing that renumbering took place was that no photographs have been seen with the 30xxx numbering, but two photographs of top door vehicles exist which are numbered 53962/4, one of which is reproduced here. Recent evidence suggests, however, that six of L507 had swing doors and four did not (as described already in this volume); moreover a drawing also suggests that

the last two vehicles of L552 had the longer wheelbase of N15 and other information that some of L507 were marked 'now on lot 552'. The simple story would be that N1/11 came from L420 and L507, that N14/15 came from L552 and that N17 came from L643 (which latter we know to be so). The situation is confused however because of the business about 'L507 now on L552' which is interpreted as meaning that some of the vehicles with L507 running numbers (i.e. 53990-99) had N14 brake details (confirmed by photograph of 53993 as N14). Another way of viewing the situation is to regard the running numbers associated with lots 507 and 552 to be different from those originally assigned, i.e. if L507 is N1/11, the numbers really were only *some* of 53990-9 and *some* of 30495-500; likewise for L552. In this way the spirit of the interpretation in Vol.1 is not wrong. However, there is insufficient information for us to determine completely which numbers went with which diagram; ironically, what information is available is conflicting. It seems that N14 vehicles were built on L507 (probably 53990-93) and when L552 was started some of these were built as N11 type to make up the original number ordered. It also seems that the six vehicles of L507 with top doors included some of the N11 type vehicles actually built later on L552 but classified under L507; the re-numbered 53962/4 would seem to be some of these vehicles, and it could very well be that the renumbering took place when the top doors were put on.

As explained by Pratt (*British Railways in the Great War*) the success of early armoured trains in South Africa and other places, led to the building early in World War I of such trains for home defence. 'Number 1 Armoured Train' had its trial run on Boxing Day 1914 between Crewe and Chester (Vol.1, p.23). The Great Northern Railway had provided a condensing 0-6-2T, the

Caledonian Railway two 30-ton boiler trucks for uns, and the GWR bogie coal wagons (53989/96) or infantry vans (garrison vehicles). All were sent o Crewe for conversion and armour plating; $\frac{1}{2}$in thick plate was used on the sides of the coal wagons and $\frac{3}{8}$in plate for a roof. Twenty-eight rifle loop holes were provided along the sides of the GWR wagons, which had lockers, water tanks and two 1-ton bunkers (!) for the troops' stove, and also as a reserve for the engine. Two more bogie coal wagons (53981/94) were sold to the War Office in 1915 for a second train. One train was stationed at Edinburgh to patrol the North British lines, and one worked the Midland & Great Northern Joint lines in Norfolk. None of these wagons returned after the war. The bogies of the remaining 23 wagons along with those off 17 J13 GANES, were given temporarily in 1918 to the strengthened 40-ton J20 military MACAWS B (later called MACAWS D), replacement bogies being built later.

The four vehicles sold to the War Office must obviously be taken into account when attempting to reconcile the N1/11/14/15 vehicle numbering. The Swindon drawing office copy of the diagram book says that in December 1939 there were six N1 vehicles, five N11, four N14, one N15 and eight N17 wagons. Note straight away that these add up to 24 not the expected 23. Two of N17 were sold in 1914-15 so N17 seems in order (having lost 53981/9) and 53996 is marked against N11 as being converted into an armoured wagon, which is also in order, but 53994 is marked against N15, which seems odd: because of the practice of numbering backwards, 53994 was built before 53993 which we know is N14. Again, 54000 had top doors yet L420 is not marked against N1 in

the Swindon book: if six vehicles from L507/552 were in fact also given top doors, N1 should have seven vehicles in total. It is very confusing. If we may be selective in what information we use, the best guess seems to be N1 54000 and six wagons from 53994-99 and 30495-8 (possibly 53994/5 and 30495-8); N11 53996-99; N14 53990-3; N15 30499/500; N17 53980-99. Note that it seems that the 30xxx vehicles were renumbered to 539xx numbers (53961-6).

Hopper wagons

The traditional GWR coaling stage, built up on an incline, involved side unloading by hand into bins that were wheeled out over the bunkers of locomotives and tipped. Hopper wagons discharging their load in bulk below the vehicle might not therefore be expected to play a part in the loco coal fleet. However the GWR had various 'industrial' plants of its own, such as the gas works at Swindon and power station at Park Royal, and these installations were equipped to receive coal on a bulk basis.

Two sizes of hopper existed (12 ton and 20 ton), and apart from evolutionary constructional changes,

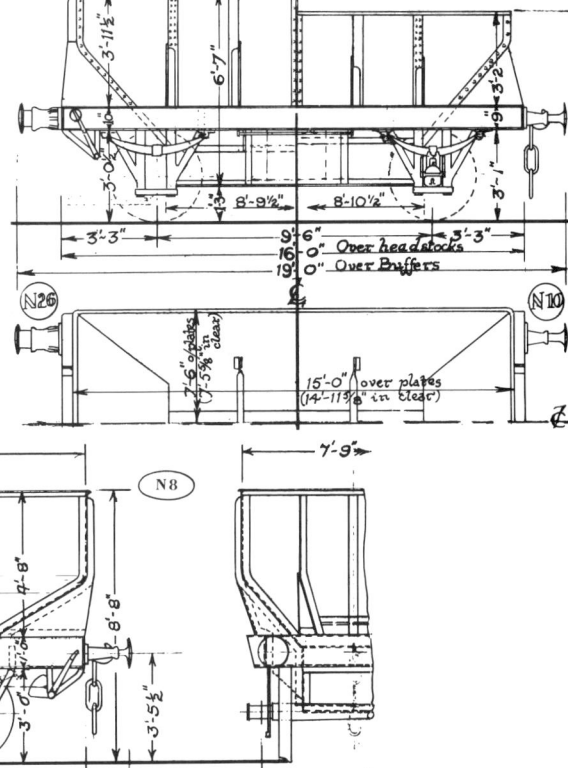

Below: *For N9, hopper base as N10/26, 8ft 4in above rail. For N12, as N9, DC III vacuum brake, 9ft above rail. For N25, as N12, GW s/c buffers 8ft 8in above rail.*

Right: *End as P7.*

Above: N12 built and photographed 1908. Based on N9. Laminated spring buffing and drawgear. 1ft channel solebar. Dust guards over springs and tops of 10 by 5in OK oilboxes. Note 3-link coupling as built even though vacuum fitted. Release 'cord' is a lever on solebar at white star above right hand spring bearer. DC III brakes, lever not painted white. Step and commode handles on either side (where hopper catch handles are) as on N25 photograph.

Left: Silt Wagon for Old Oak Common drains photographed in May 1936 when converted from 10ton N13 loco coal wagon No 9068 built 1906. 16inch lettering. Classic Churchward 9ft wheelbase iron channel underframe. Tucked-under side T stanchions, destination box, single-ended DC I brake, note inverted brake shafts on rocking shaft; brake blocks (by 1936 reversible) this side only. Vertical rod inside V-hanger. 9in by $3\frac{1}{4}$in GW oil boxes.
(Late E. McDougal Collection)

Left: N25 built 1925, photographed 1938 after vacuum brake removed. Based on N12. DC cross-cornered brake (white stripe). 10 by 5in OK oil boxes. 1ft channel solebar. Rectangular works numberplate. GW self-contained buffers. 'Loco' here means locomotive department, not for loco coal, since legend says 'EMPTY to Markham Colliery Staveley Town. LOADED to G.W. Gas Works Swindon via Bordesley Junction'. Note catch handles and locking 'french pins' to operate hopper.

there were only two designs. Two diagrams were issued for 12-ton hoppers, N10/26. N10 and the ballast hopper P7 were contemporary wagons dating from 1893. All P7 were reconstructed to 20-ton vacuum-fitted wagons after 1902, but the loco coal hoppers remained at 12 tons in new construction. However, whereas the original N10/P7 design had a hopper some 2ft 9in above the solebars with only a small vertical section, L527 of 1906 and L637 of 1910 (43884-6 for Swindon Carriage Works generating station) built N10 at 3ft 2in above the solebars, the amount of top vertical plating being increased. Strictly therefore diagram N10 does not apply to the 70 loco hopper wagons of P7 type built during 1893-8, and those wagons were really omitted in the original listing, as their logical place is lower in the list. Following the 13 wagons of N10, 25 12-ton wagons were ordered under L908 in 1923. Given diagram N26, their body followed the 'one-off' P6 of 1902, with yet higher flat portions at the top of the hopper, which was now 3ft $11\frac{1}{2}$in above the solebars. The underframe contained later features such as GWR self-contained buffers. Oddly the brake system remained at DC II (cf N26). In common with other building times in the mid-twenties the lot took three years to finish.

The first two large hoppers on the GWR were N8/9 ordered in 1902. Heavy 1ft channel was used for the 21ft over headstocks 13ft wheelbase underframe, a practice which continued on the later wagons, N12 of 1906 and N25 of 1923, until 66 20-ton hoppers finally were built. Apart from N8, which had a different underwagon discharging system, all the bodies were the same in general layout; even N8 was like the others above the solebars. The differences between the diagrams relate to hopper height, as with the 12-ton wagons. N8 was 8ft 8in above rail level, N9 8ft 4in, N12 9ft, and N25 8ft 8in. Following the success of N9 principally, for it was this wagon that had conventional hopper doors the length of the vehicle, N12 built 39 wagons in 1906-7. As constructed with DC II brakes, they were vacuum-fitted and eventually had instanter couplings (cf N2), though these were later removed and three-link couplings substituted; 53082 amongst others worked the Park Royal electric station traffic. The parenthood of the 25 wagons of N25 is evident from one of the drawing numbers which marks them as 'revised' N9. GWR self-contained buffers and drawgear, DC cross-cornered brakes and Gedge hooks were underframe changes, Vol.1, p.51. The long building time of N26 is seen here too, since N25 was not completed until 1927. Many of these wagons were used in the Swindon Gas Works coal and coke traffic.

To achieve a load discharge centrally between the wheels, the point of the hoppers was some 5in to one side of the wagon. The hopper door was worked by two pairs of catch handles connected by $1\frac{1}{4}$in gas pipe and locked by 'french' pins. These arrangements were typical of ballast hoppers too.

In capacity, height and width, a definite trend can be seen in the coal wagon fleet: N4 possessed 30cu ft of space per ton of coal; N3/7 had 35cu ft available, and all the remaining wagons had about 40cu ft per ton. This was the figure allowed by the GWR after the turn of the century for Welsh coal; the RCH allowance was 44.5cu ft per ton. This suggests that in the early days coal wagons 'rode high' (see BGW p.26) and offers an explanation for the extra rails on N18, since it was likely that without them coal would fall over the back of the wagon when tipped. When some of the 20-ton wagons were uprated by one ton in World War II, presumably they too showed their loads more. Curiously, the trend in ballast wagons was the other way, the early ones allowing 25 or 26cu ft per ton of ballast, and the later ones the regulation 21cu ft per ton. Other allowances were 73cu ft per ton for foundry coke and 21cu ft per ton for iron ore.

CHAPTER 13
O – OPEN MERCHANDISE WAGONS
(Opens)

The ubiquitous open wagon was used as an example to trace the constructional details of GW vehicles in Volume 1. One- and two-plank open wagons with round ends carrying 8 tons were common before 1870, and were called 'low-sided' trucks. The last two-plankers, 15ft 6in over headstocks 11in inside height, were built under osL54 in April 1872 and numbered in the 15xxx and 21xxx series. The first three-plankers, 1ft 9in inside height rated at 9 tons, had just been ordered under osL40/1 at Saltney and Worcester, in the extremely large batches for those days of 500 each. They were numbered in the 18xxx series. Round ends gave way to square ends on osL289 in 1883, and by 1885 or so nearly 7000 3-plankers had been constructed, including some broad-gauge convertible wagons and 50 with iron bodies (osL348).

Most of these wagons had wooden solebars, Vol.1, p.53 (and dumb buffers on the earliest), but flitched frames appeared on the 200 wagons of osL211 in 1880 numbered 31401-600. This led to the introduction of iron underframes (at first bulb section), and then all the standard four-

plankers built between 1886-1902 had channel frames (cf contemporary building of V6 iron MINKS). The length over headstocks was increased to 16ft on these 10-ton wagons, and their inside height was 2ft 4in. Some experimental 18ft four-plank 12-ton wagons were built in 1887-8 (osL367/454), numbered 42896-42900, to test the possibilities of relatively low tare larger capacity wagons, but for reasons explained later in this Chapter in connexion with the TOURN, the idea was not followed up immediately.

Those four-plankers fitted with the DCI brake (two of L355 and all of L374) were given the O5 diagram when the index was set up, but *none* of the single lever brake one-, two-, three-, or four-plank OPENS was put in the index. There were about 2000 single plankers still in service during World War I; as late as 1934 some were refurbished into match trucks (L21 diagram). From 1927 over 18,000 four-plankers were given an additional lever brake to comply with ministry 'either-side' regulations; this was the orgin of the O21 diagram. It should be emphasised that prior to this, only

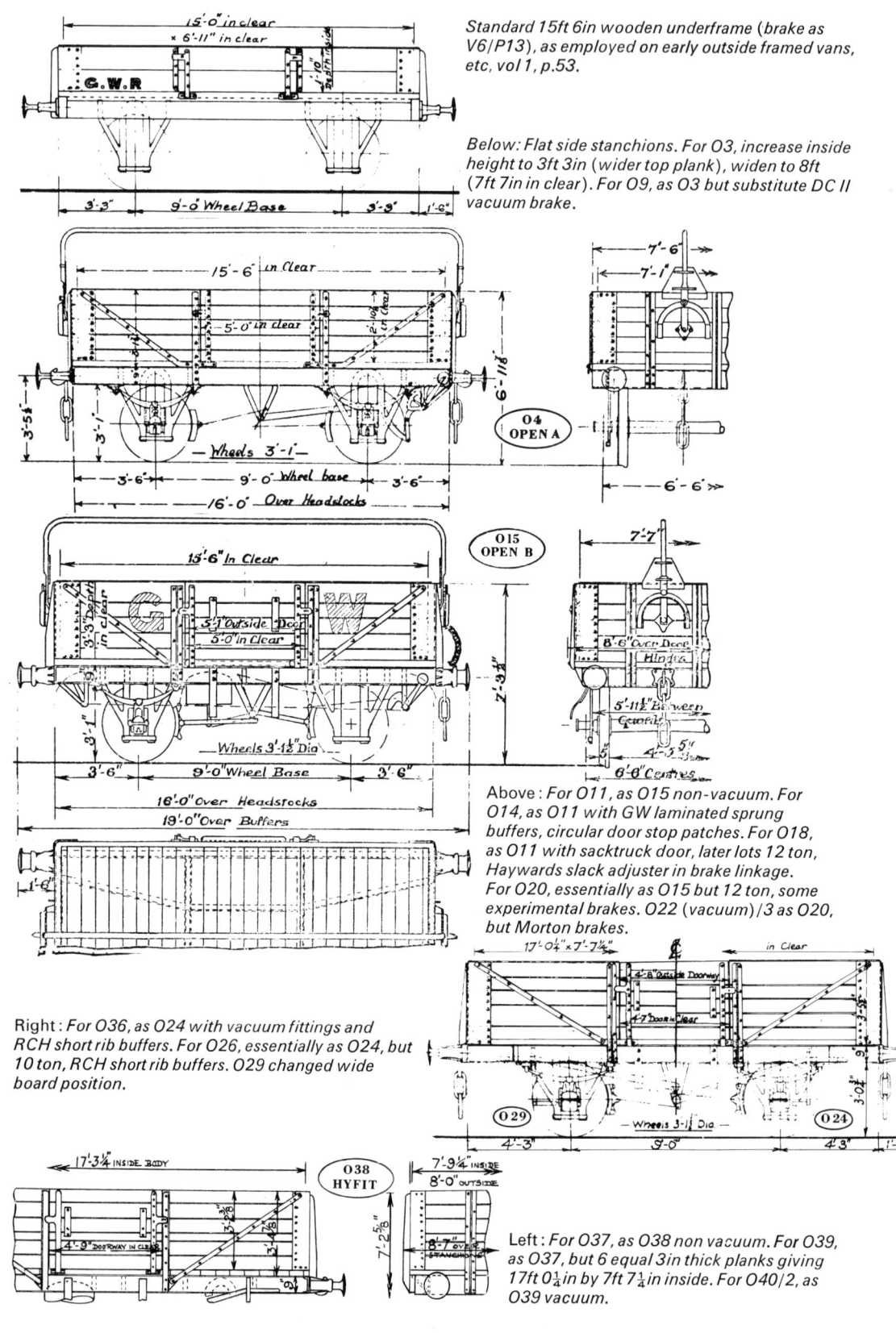

Standard 15ft 6in wooden underframe (brake as V6/P13), as employed on early outside framed vans, etc, vol 1, p.53.

Below: Flat side stanchions. For O3, increase inside height to 3ft 3in (wider top plank), widen to 8ft (7ft 7in in clear). For O9, as O3 but substitute DC II vacuum brake.

Above: For O11, as O15 non-vacuum. For O14, as O11 with GW laminated sprung buffers, circular door stop patches. For O18, as O11 with sacktruck door, later lots 12 ton, Haywards slack adjuster in brake linkage. For O20, essentially as O15 but 12 ton, some experimental brakes. O22 (vacuum)/3 as O20, but Morton brakes.

Right: For O36, as O24 with vacuum fittings and RCH short rib buffers. For O26, essentially as O24, but 10 ton, RCH short rib buffers. O29 changed wide board position.

Left: For O37, as O38 non vacuum. For O39, as O37, but 6 equal 3in thick planks giving 17ft 0¼in by 7ft 7¼in inside. For O40/2, as O39 vacuum.

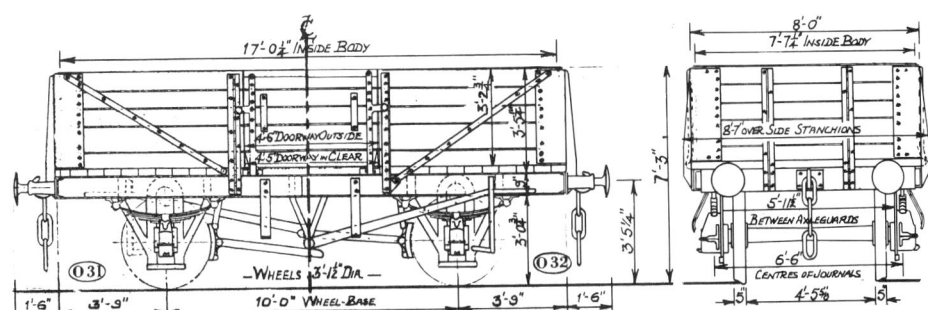

the DC-braked O5 were in the index. Reverting to the development of the open goods wagon, a fifth plank was added in 1902-4, giving 2ft 11in inside (O4 Vol.1, p.46), and also the first Williams patent sheet supporters; before this, tarpaulins were simply draped over wagons, and the problems caused by such hollow sheeting have been mentioned in Volume 1. The Marillier patent sheet supporter was tried on an undiagrammed four-planker (RWA Fig 43), but unlike his cattle truck partition device (with F. G. Wright) it was not adopted. Some four-plankers received supporters later. In 1904 the top fifth plank was widened to

11in (O3) giving a 3ft 3in inside height, which then remained basically the standard height through the life of the GWR on 10- (later 12- and 13-) ton wagons; at the same time the width of the wagon was increased to 7ft 7in (8ft outside). Sloping bottom planks in the doors (tapered feet) were introduced at the end of World War I to facilitate loading with sack trucks when the door was dropped on to a platform. Some time after the grouping, open wagons were built on RCH 17ft 6in underframe with detail differences. A summary of the nine 16ft over headstocks 5-plank diagrams that were incorporated in the index and the twelve

Below: *Three-plank 8 ton open. Built circa 1875 or earlier. Photographed 1890s (?) Flitched solebar, 10 leaf springs. Mixture of odd wheels (open spoke and solid) and buffers (ribbed and plain early types). Note attachment coupling to hook. Timber end stanchions. Label 'frame' below number. No 'Tare' with 5–3–0. Vehicle behind is end of provender wagon (Q1), with end L-stanchions.*

Above: *For O33, as O32 with vacuum brake, instanter couplings and strip tiebars between axleguards.*

Above : *O9 OPEN B
photographed 1909. First 16ft
opens with DC III brakes,
vacuum, 4 blocks, tiebars.
Instanter coupling,
non-Gedge hook. Laminated
sprung buffing and drawgear.
White label box.*

Left : *O22 OPEN A built and
photographed 1925. GW
body and u/f (O18-type), but
Morton brake. Pinned guide,
2 brakeblocks. Straight
diagonals, GW s/c buffers.
Not Common User plate
(note non-vacuum
nevertheless). Sacktruck
sloping door, hinges curl
under. Black quarter top right
hand side. Label clip near
GW rectangular works plate.*

Left : *O24 OPEN introduced
1924, photographed 1925.
First 17ft 6in oh, 10ft wb
RCH design (same u/f details
on V21/33 and W11). RCH
long rib buffer ('GWR' cast
on base), RCH 9 by 4½in
boxes, Morton brake, pinned
rack, 2 shoes. Corner plates
cover curb rail (unusual).
'Oval' works plate. Horse hook
(GWR holes previously),
label clip on block of wood
over axle.*

TABLE 3   16ft over headstocks wagons

Diagram	Building Period	Tonnage as Built	Vac ?	Brake	Buffers	Side Diagonals	Door	Sheet Supporter	Original Coding
03	1904–5 & 1912	10	both	DC I & II	laminated sprung	curved feet	flat with lip	yes	OPEN A & B
09	1906–11	"	nearly all vac	DC II	"	"	"	"	mostly OPEN B
011	1911–19	"	non vac	DC III	mostly GW self contained		"	"	OPEN A
014	1910–12	"	"	"	laminated sprung	curved feet	"	"	"
015	1907 1911–12	"	vac	"	GW self contained	straight	"	"	OPEN B
018	1919–24	10 & 12	non vac	"	"	"	tapered foot	mostly all	OPEN & OPEN A
022	1924–5	12	both	Morton (some DC III)	"	"	"	"	OPEN A & B
020	1925–6	"	vac	DC III	"	"	flat with lip	"	OPEN B
023	1926–7	"	non vac	Morton	"	"	tapered foot	no	OPEN
025	1925–6	"	"	"	"	"	"	yes	"

All 16ft wagons were 15ft 6in x 7ft 7in x 3ft 3in inside, with 5ft in clear doors and the wide sideplank topmost. All had a 2¼in curb rail covering the edges of the floorboards. The side planks were 2¼in thick and the ends 3in. On 03/9 the end stanchions were 'parallel' T-irons, and the side stanchions flat; on all the others both end and sides had tapered T-stanchions. The side stanchions on all 'tucked under' the wagon to the solebars. The hinge straps themselves acted as doorstops on 03/9, then separate strap patches or discs were used in the middle of the door. 'Sacktruck' doors (tapered foot, with bottom plank angled out) appeared after World War I.

17ft 6in diagrams is given in Tables 3 and 4. Note that the captions on pp.49/50 of Vol.1 of O32 and O29 are erroneously inverted. On those pages the development of OPEN design changes is discussed. Most 12-ton wagons were uprated to 13-ton in World War II. The 1939 shock absorbing OPENS were not incorporated into the index as O44 until after nationalisation.

Most wooden opens were built at Swindon, though some came from outside builders. For example in 1912 the Metropolitan Amalgamated Wagon and Finance Co supplied some of O11, and

immediately after World War I 97001-500 of the first 12-ton OPENS A (O18) were built at the Royal Arsenal for the GWR under war production measures. The pattern wagon sent to General Martell at Woolwich in November 1919 (Order O.1423) was 94430 of L850. Orders 1490/3 of March 1921 were issued to cover the expense of making adjustments to open wagons delivered by some other outside commercial builders.

Seven-plank wagons, 4ft 3in deep, were built during 1905-7; O10 was the vacuum version of O2 (DC II brakes instead of I), with also some detail

O44 open shock absorbing wagon. Built 1939 (no diagram until BR), photographed 1964. RCH 17ft 6in u/f and fittings, shortened 16ft 6in O38-type body.

(W. Beard)

O10 OPEN B built 1906, photographed 1920, painted with 25in letters. Seven-plank, 4ft 3in deep. Extended sheet supporter, vacuum fitted, 4 blocks, tie bars, white star for release cord, though 3-link coupling non-Gedge hook. Cylinder on far side. Single-ended DC II brakes (left handed this side). Side drop door similar to ordinary opens, circular door stop patches; swing doors 3 planks deep, often only 2 planks on others. Lashing rings left half only. Black patch top right.

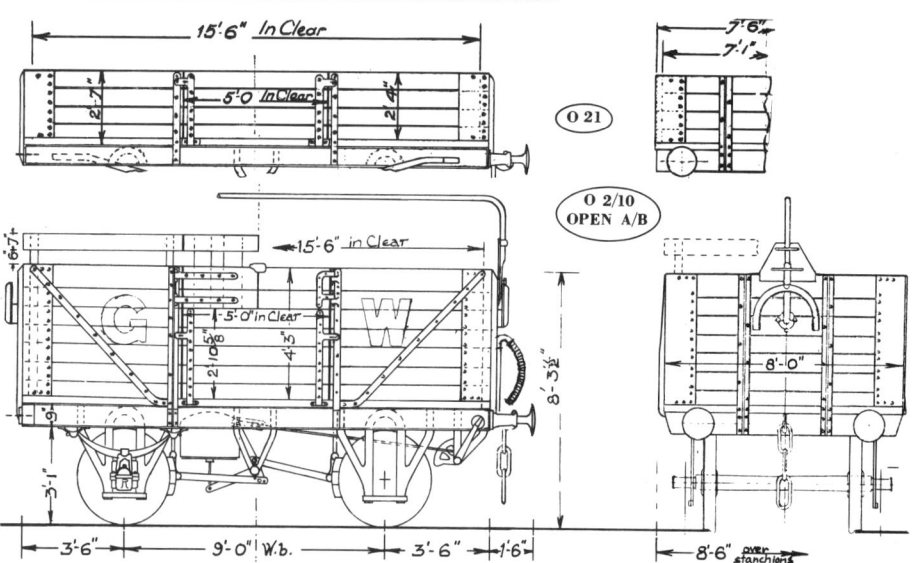

Top: Underframe as P13/V6, but with a separate lever brake each side, ie one additional to original condition. For O5, body as O21; DC I u/f as O4.

Above: For O2, underframe as O4, body as right portion (flat side stanchions etc). Top rails for World War I military horse traffic (bottom left stencilled 'For military use only').

Below: For O16, body as O19, u/f as O8 without chain boxes. For O28, body as O19 (thinner planks to give 7ft 7¼in in clear), no taper on side stanchions, RCH buffers. For O34, as O28, but 'straight-down' angles, exposed floorboards, 9in corner plates RCH axle boxes.

Right: O8/16 OPEN C built and photographed 1914. Curved feet to diagonals of O8, but absence of chain boxes and presence of lashing rings suggest O16. DC III brakes central white stripe, handles not painted white.

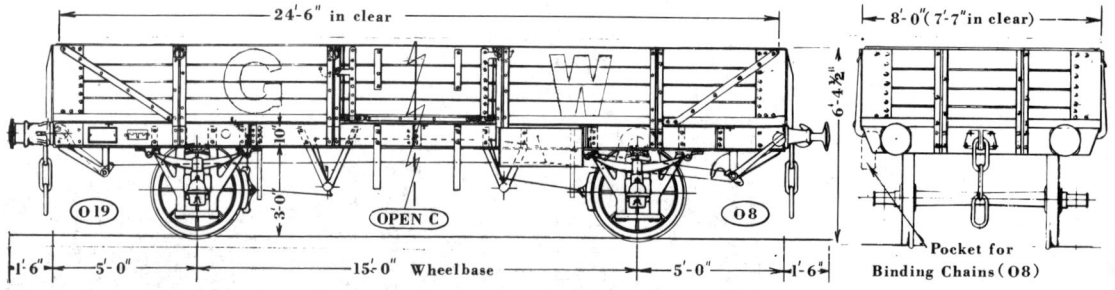

TABLE 4   17ft 6in over headstocks wagons

Diagram	Building Period	Tonnage as Built	Vac ?	Brake	Buffers	Side Diagonals	Door	Sheet Supporter	Original Coding
024	1925-9	12	non vac	Morton	RCH self contained	straight	tapered foot	no	OPEN
036	1926-8	"	vac	Morton (some DC III)	"	"	"	yes	OPEN B
026	1928	10	non vac	Morton	"	"	"	no	OPEN
029	1930-4	12	"	"	"	"	"	"	OPEN
031	1932-3	"	"	"	"	"	"	"	OPEN
032	1933-40	12 (13 after 1937)	"	"	"	straight gusset foot	"	"	OPEN
033	"	"	vac	"	"	"	"	yes	OPEN B
037	1939-45	13	non vac	"	"	"	"	no	OPEN
038	1944-7	"	vac	"	"	"	"	"	HYFIT
039	1945-BR	"	non vac	"	"	"	"	"	HIGH
040	1945-BR	"	vac	"	"	"	"	"	HYFIT
042	1945-BR	"	"	"	"	"	"	"	HYFIT

The 17ft 6in wagons came in two groups: (i) the early 9ft wheelbase wagons 024/36/26/9 which were 17ft x 3ft 3in inside with 4ft 7in clear doors and which were merely lengthened versions of the 16ft wagons with slightly thinner planking, and (ii) the RCH 10ft wheelbase vehicles. The first of these, 031, was erroneously sent through the shops as 3ft 2¼in inside height instead of 3ft 3in, nevertheless the shorter size perpetuated on all further designs to nationalisation, except for the Wartime 037/8 which used a selection of six 1⅜in thick planks for sides and ends which gave 3ft 2⅜in inside height.   031 had a 4ft 5in in clear door, but all others became widened to 4ft 9in when L-irons replaced T.   The width over stanchions, for so long 8ft 6in, became 8ft 7in on 032 et seq when the stanchions went 'straight-down', the side diagonals became attached with a gusset foot, and the floorboards became exposed.

Door stops on the 17ft 6in wagons were mostly separate strap patches, - wider apart than the 16ft wagons.   The deep side board took a new position on 029, and became second up from the bottom.   After 1936, the corner plates were reduced to 9 x 9in from the traditional 12 x 12in.

changes in the body such as the method of hinging the top swing doors. The lower part of the door was essentially a five-planker type of drop door. With the supporter bar raised, these wagons cleared nearly 11ft above railhead. During World War I, the sheet supporters were removed from 410 seven-plankers and top rails were fitted, along with slots in the floorboards, for the conveyance of military horses within Great Britain. Although this contravened a 1904 Board of Agriculture regulation forbidding open animal trucks, it was allowed during the war; 28023 was one such vehicle (RWA Fig 38).

Diagram O8 concerned the first of the well-known OPENS C; 25ft over headstocks, 15ft wheelbase with a 14-ton load, these wagons were intended for deals of timber. Later steel tubes became common loads, and it was to TUBE that the vehicles were recoded in World War II, when they were also uprated to 15 tons. The first lots in 1907 had chain and shackle boxes, but subsequent wagons had merely lashing rings in the top of the

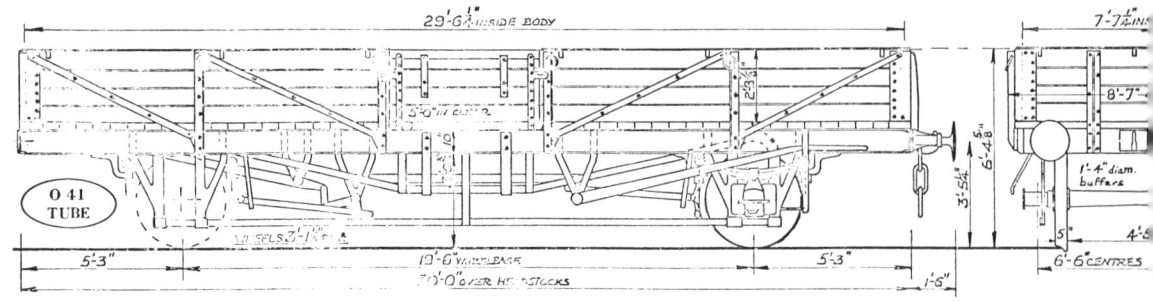

Above: *O41 TUBE, the post second world war 21-ton design with 19ft 6in wheelbase and length over headstocks of 30ft. It was fitted with the Morton handbrake, modified for the long wheelbase.*

TABLE 5

Side body diagonals	Only 08 had curved feet
Side stanchions	T-section used on all except 034, which had L-section
Door design	The 'sacktruck' door with tapered foot appeared on 028/34, the rest had flat doors with metal lips
Buffing and drawgear	08/16 had laminated spring equipment, 019 had GWR self-contained buffers and Gedge hooks, and 028/34 had RCH fittings
Floorboards and corner plates	034 was the only series with exposed floorboards and 9in by 9in corner plates

Note: The dates of building show that the OPENS C were late in getting many new pieces of equipment compared with most 16ft or 17ft 6in open wagons.

side stanchions. The OPENS C were among the first wagons designed new with the DC III brake. Detail differences between the five diagrams, O8/16/19/28/34, shown in Table 5, reflect typical constructional fashions.

In 1945 a new 21-ton TUBE design was introduced, 30ft over headstocks, 19ft 6in wheelbase, which became O41. One hundred of these wagons were built into nationalisation. The long wheelbase modification of the Morton brake was a feature of the design.

Below: *O34 OPEN C built and photographed 1937. Exposed floorboards, small 9in corner plates. Straight down L-stanchions; diagonals attached without gusset plate since outer stanchions are 'turned about'. Short rib buffers DC III brake (Haywards slack adjuster)—4 blocks 8 by $4\frac{1}{2}$ in OK boxes.*

Above: *O1 TOURN photographed when built in 1889 (by 1891 was a 4-planker). Corner hung 4ft 10in wb bogies, grease boxes, lever brake other side. Circular headed buffers. Fixed kingposts.*

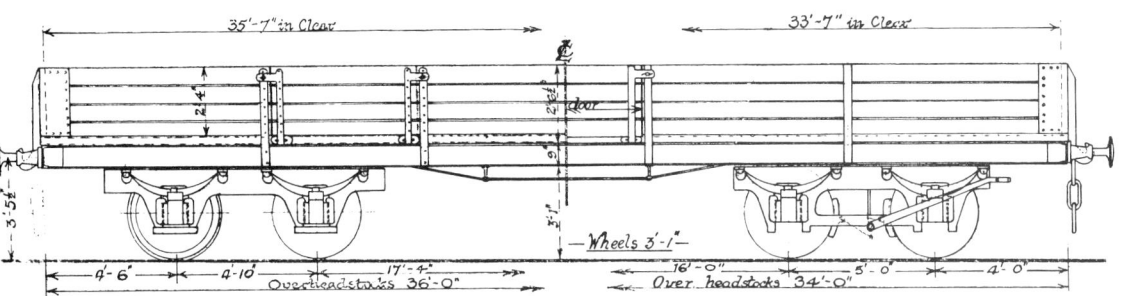

O1 Tourn

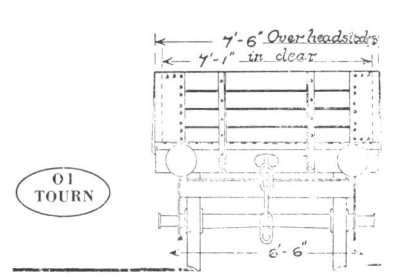

42902 was a 25-ton bogie open built in 1889 at the time when large-capacity foreign wagons were attracting attention in Great Britain. J. L. Wilkinson had just returned from Argentina, where he had been General Manager of the Buenos Aires & Pacific Railway, to rejoin the GWR in the post of the Chief Goods Manager. There is confusion over the dimensions of this wagon. It seems that originally it was intended to build it 34ft over headstocks, with 5ft bogies, on an identical underframe to the contemporary P1 bogie ballast wagons. All published sketches, however, give the length as 36ft over headstocks with 4ft 10in bogies. Again, the lots state an inside height of 2ft 4in (four planks), yet the 1889 *Special Wagon Book* gives the inside height as 1ft 3in, the 1893 edition gives 2ft 6½in (confirmed in the 1900 telegraph code book) and the 1910 edition 2ft 4in. All show a single 5ft wide central door. Clearly the photograph demonstrates that it was built as a two-planker, but it was upon inside dimensions of 35ft 6in by 7ft 1in by 2ft 4in that Dean, in a letter to the Locomotive, Carriage and Stores Committee at Paddington, dated 27 January 1891, could quote a capacity of 600cu ft. Thus two extra planks were added within a year or so of building. The compound drawing given here is based on two

Swindon drawings, one of the 34ft version (which may never have been built) and one of the 36ft version, but with the two doors per side which it had after March 1922 (NWO.F419).

The business of the country was dealt with in small lots, the railway rating system giving no incentive for bulk loading in general merchandise traffic, and it was difficult to find full loads for the TOURN, even between large centres such as London, Birmingham and Bristol. No other large-capacity open merchandise wagons were built, and 12 tons was chosen at the RCH standard after amalgamation, even though the average wagon load in the 1920s was about 3½ tons; if anything, this figure was falling because of the insistence on quick transit in reply to road competition.

Five further bogie open goods wagons were ordered under L402 in 1903 (42903-7) but were

O7 linoleum wagon built 1890, photographed about same time. The first 5-plank open built by GW. Round ends in lieu of sheet supporter. 'Short' doorstops (square patches on door). The second wagon ever to be fitted with oilboxes (OK type). 4ft 6in springs 'Γ'-hangers. One-sided lever brake far side. Blocks swung rigidly.

O35 photographed when built August 1939. One of two pattern wagons for DX containers (Bath and Portland stone). First new 3-planker since 1885, and only type to use side diagonals. RCH 10ft + 3ft 9in vacuum u/f with tiebars (as contemporary O33), but 1ft 8½in buffers, screw couplings and lamp irons. 9 by 4¼in oilboxes, Morton brake, ratchet guides. Exposed floorboards, small 9in corner plates. Straight down L-stanchions, diagonals attached to gusset plate. Writing as on CONFLATS.

O25 open goods wagon convertible for grain traffic (Barnard's patent). Built and photographed 1926. GW 16ft body (Morton brake as O22, 2 blocks) with hopper in floor. Note handwheel. Trap doors over hopper swing up against side doors when carrying bulk grain. RCH 9 by 4¼in boxes.

O30 steel-bodied open goods. Photographed when built 1934. Wood floor, woodlined doors to prevent slipping when loading. Flat door (not sacktruck), lip at bottom. Hinge straps act against door stops. L-angle diagonal bracing. 'Straight down' L-stanchions, gusset plate attachment of diagonals. Corner plates also straight down over buffer beam (ends of which are not chamfered). U/f details as O24 except RCH short rib buffers and toothed brake rack. Label clip fixed directly to solebar, extreme left hand. No number on end (had ceased on O29).

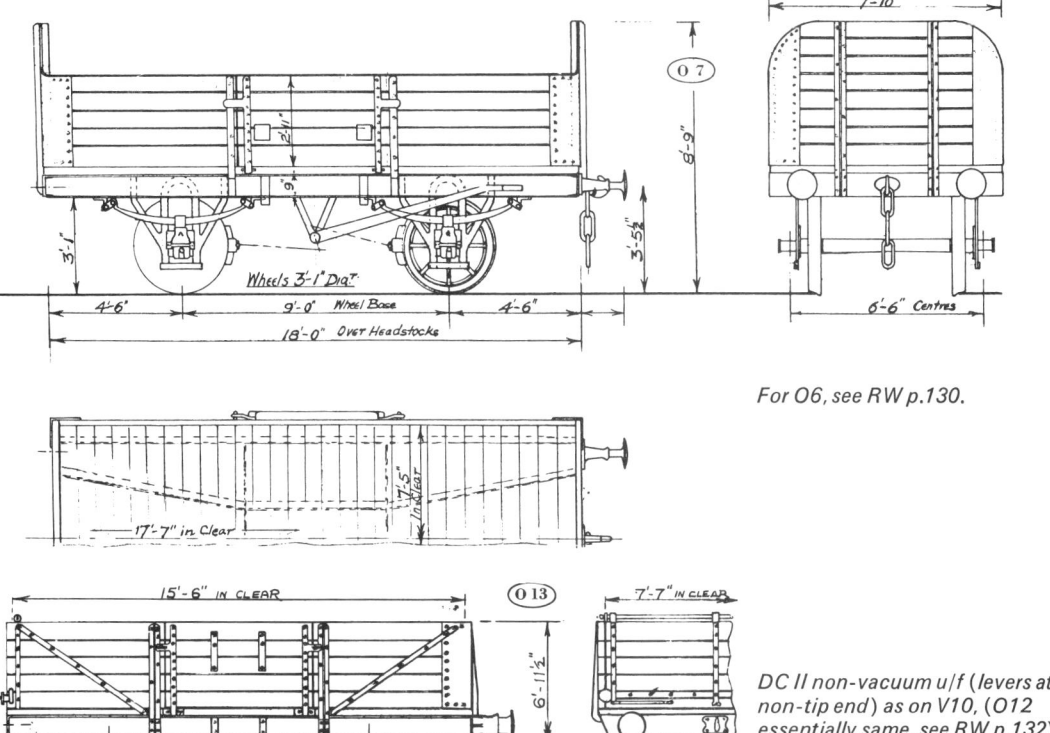

For O6, see RW p.130.

DC II non-vacuum u/f (levers at
non-tip end) as on V10, (O12
essentially same, see RW p.132).

cancelled. It might be thought that the 34ft over headstocks drawing relates to these wagons, but the bogies were not the plateframe design that would have been logical in 1903. The TOURN was scrapped in the summer of 1934.

O6/7 linoleum wagons

Two vehicles were built for this traffic, O7 in 1890 and O6 in 1905. O7 was one of the first goods wagons to be fitted with oil axleboxes; it was wider at 7ft 10in than contemporary OPENS and its five-plank body presaged O4. Both vehicles, 18ft over headstocks, had long springs, and were characterised by high round ends (used by other companies in lieu of tarpaulin supports, although O6 had a sheet supporter as well). In 1908 the four-plank O6 was converted to a five-plank wagon, and also received (rather late it seems!) a DC I brake; before scrapping in 1942 it had been converted to square ends.

O12/13 china clay wagons

A fleet of wagons was in existence before World War I for the carriage of china clay from the Fowey branch. They were essentially contemporary open designs (O11/14/15), but with five equal 7in planks with an end tipping door; furthermore, the floors

(which had longitudinal planking) were lined with zinc to prevent staining the clay. A pattern 10-ton wagon was built in 1910 (L656, 42833) and 25 10-ton vehicles were purchased from outside contractors (O12). Then after the proof of the design, 500 more wagons were built at Swindon in 1913. These were 12-tonners (O13), using 8in by 4in journals rather than 8in by 3¾in, and in the late 1920s O12 were uprated to conform. All wagons had DC II brakes, which was slightly out of date, but presumably their use avoided the mechanism being gummed-up at the tipping end. After 1939 they were given Morton brakes to conform with the 'righthanded' ministry rulings. The wagons saw hard service in World War I when much china clay traffic was diverted from coastal shipping because of the shortage of ships and the submarine danger.

As regards capacity allowance, china clay took about 28cu ft of space per ton.

O17 wagons

The original holders of this diagram were 400 'high-sided wagons with roof doors for overseas', which were O11 wagons converted under Order O.748 in April 1917. They looked like the pitched-roof salt wagons used by private owners (RWA Fig 44). Those that were returned reverted to O11.

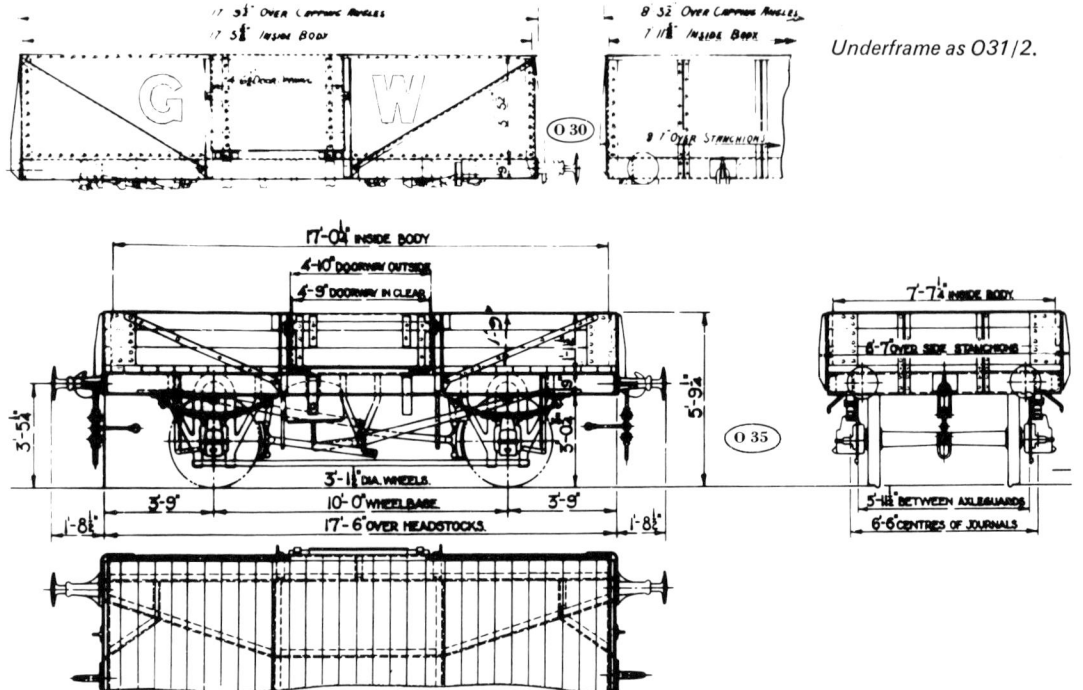

Underframe as O31/2.

At amalgamation, 28 open wagons were built from details and materials supplied by the Cambrian Railways in 1922, and they took over the vacated diagram. They were non-standard 17ft over headstocks, 9ft 6in wheelbase five-plankers with an inside height of 3ft; Cambrian pattern self-contained buffing and drawgear was employed. These wagons had the distinction of being the first Morton-braked wagons on the GWR, unless the single nine-plank coal wagon (L525) purchased from Gloucester RCW was so fitted.

O25 12-ton OPEN A convertible for grain traffic

In 1925 six standard 12-ton non-vacuum open merchandise wagons were built with hoppers below the floors to allow discharge over staithes. Trap-doors were provided which swung up against the regular side doors when used for grain, but which could swing down to cut off the hopper and allow the wagons to be used as normal OPENS A (Barnard's 1919 Patent). The convertible concept had been applied in vans with roof doors (V10 in 1905), and shortly was used again in V20.

Evidently the experiment was not successful, for the hopper was removed on 109438/40/2 in the 1930s and the wagons returned to normal service; the other wagons 109437/9/41 remained as O25, but even though the hopper was left, the floor was

fixed down. V20 were converted to cement wagons (V29) at the same time, the importation of grain for which they were built having temporarily ceased.

O27 20-ton merchandise wagons for tinplate bar traffic

Three vehicles were built in 1931 for the heavy traffic passing between steel and tinplate works in the Swansea district. The aim was to introduce a wagon of higher carrying capacity than the usual 10 or 12 tons but not longer than the standard 12-ton wagon, in order to negotiate restricted private sidings. They were therefore similar in outward appearance and design to ordinary open wagons with wood bodies, but the metal underframe was of 10in channel for the heavier load. To protect the inside of the body at the ends, plates 1ft 6in deep and $\frac{1}{2}$in thick were fitted to the end sheeting for the full width of the wagon. Although RCH buffing and drawgear were used, DC III brakes were employed. They were 16ft 6in over headstocks, 1ft shorter than the 'new' RCH open bodies, and had 9ft wheelbase with 3ft 9in end overhangs. Width over body was 8ft and depth of body inside 2ft 11in. They were 'Not Common User'. A new 'series' of these wagons is mentioned by O. S. Nock *History of the Great Western Railway* Vol. 3), although no more were built.

O30 12-ton open goods wagons with metal bodies

Fifty standard 17ft 6in RCH underframes were equipped with open bodies made of steel in 1934. The floors were wood however (in contrast to most mineral wagons) and the doors were lined with wood on the inside to prevent slipping when loading. Dimensions were similar to contemporary OPENS.

O35 13-ton open goods wagon for container traffic

The 'D' containers were large open types specially constructed for carrying baths; a low-sided variant of the basic design was coded 'DX' and was suitable for loading Bath and Portland stone. It was for these DX containers that two pattern three-plank opens were built in 1939 with special screw shackles. The vacuum underframe and fittings were as in the contemporary O33, but 1ft 8½in buffers and screw couplings were employed. No merchandise three-plankers had been built since the 1880s and the use of side diagonal strapping on such low wagons was a new departure. Since these two wagons were the only three-plankers in non-service use by World War II, they were the only GWR wagons to be coded MEDFIT.

Other wooden open wagons, not in merchandise traffic, were included in the J, Q, R, T and FF listings.

CHAPTER 14
P – BALLAST AND SAND WAGONS

Railway ballast, whether stone, broken slag or cinders, was traditionally conveyed in wagons holding about 6 tons and shovelled out into the six-foot, whence at a convenient time it was placed into the track. Towards the end of the nineteenth century various companies began to use hopper wagons to convey the ballast from the stone-breakers to the site and there rapidly to unload it directly into the track. Rodger's New Zealand patented method additionally involved the use of 'spreaders' or ploughs at the tail of the hopper train to distribute the ballast evenly (see AA later). The GWR was the first railway in Great Britain to use this idea, and found it of the greatest service in the re-ballasting of the line when transverse sleepers were replacing longitudinal baulks after the abolition of the broad gauge. Two men in ten minutes could do the work which by the old method would take a gang of 30 or 40 men some hours.

The hopper wagons built for this work became P7 in the index. They were designed and built at the same time as the 12-ton loco coal hoppers, N10. Some 420 were built in the period 1893-1901. The earliest lots had grease axleboxes, plate covers over the axleguards, one-side lever brakes, and low profile hoppers with hardly any flat side; from L181 in 1897 oil boxes were fitted and some had the Thomas brake. Between 1902 and 1906 extensive reconstruction took place whereby the wagons were first increased in capacity to 20 tons and then converted to automatic vacuum and DC brake. The pattern for these changes was P6, which was a 'one-off' 20-ton ballast hopper built new in 1902. A detailed drawing of the wagon is given in Vol.1, p.63. Interestingly P6 had a cross-cornered version of the DC I brake. The converted P7 attained the increased height of P6 by adding 1ft high flat sides around the top of the hopper; there was scant room for the vacuum cylinder, which hung out from under the solebar. L354 of 1901 is interesting, in that it built one wagon 60716 to replace the same

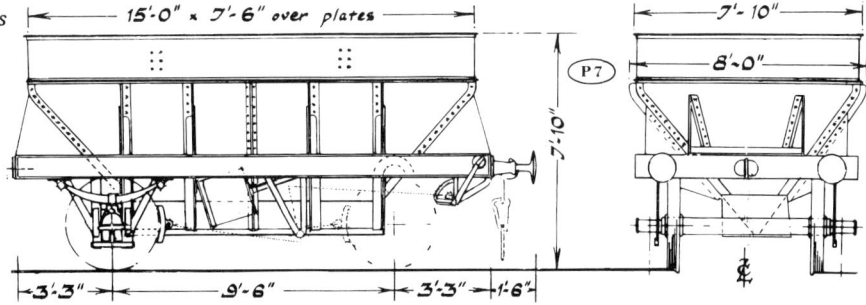

For P6, one-piece sides as N26, 10in channel u/f, originally DC I cross-cornered brake. See photograph page 76.

15'-0" × 7'-6" over plates   7'-10"

8'-0"

P7

7'-10"

3'-3"   9'-6"   3'-3"   1'-6"

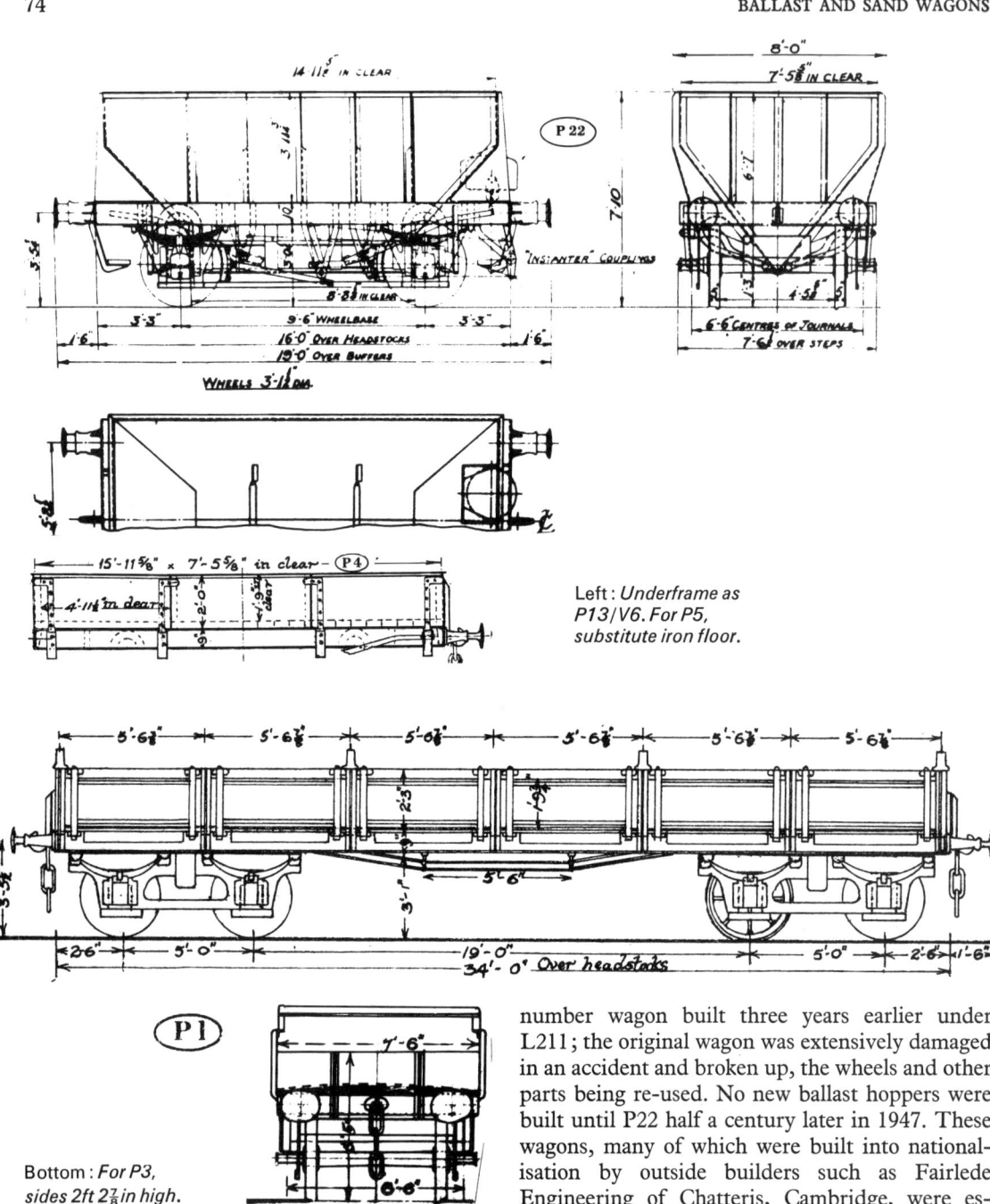

Left: *Underframe as P13/V6. For P5, substitute iron floor.*

Bottom: *For P3, sides 2ft 2⅞ in high.*

number wagon built three years earlier under L211; the original wagon was extensively damaged in an accident and broken up, the wheels and other parts being re-used. No new ballast hoppers were built until P22 half a century later in 1947. These wagons, many of which were built into nationalisation by outside builders such as Fairlede Engineering of Chatteris, Cambridge, were essentially updated welded P6.

Details of the hopper release mechanism may be found in the N section on loco coal hoppers.

There were of course ordinary open ballast wagons in the early days; some four-plankers were fitted with double floors for engineering department stone traffic but no new wooden ballast wagons were built after Churchward reigned in the carriage and wagon shops in the 1880s and iron ballast wagons paralleled loco coal wagons and

MINKS. Old wooden wagons served as 'fill' wagons for the construction department. Ballast wagons always had square corners, because of the fall-down sides. P4/5 were 8-ton (7cu yd) wagons, 1ft 9in high, built on the simple 16ft N6/7 type underframe. The sides were divided into a central fixed 5ft 6in panel, and two 5ft 3in end panels having 4ft 11½in drop doors. P4 had a wooden floor, P5 a steel floor and was thus 3in lower in height above railhead (Vol.1, p.51). Some 450 of these wagons were built in the 1890s under both diagrams and there were many other similar wagons dating from the broad-gauge era. P5 had the distinction of being L1 in the new series.

After this time, the 10-ton wagon (8cu yd) was the smallest manufactured. P3 of 1896 was the first of this size and was the same 7ft 6in width and 16ft length of P4, but gained extra capacity from the 2ft 3in sides. The three equal 5ft 4in side panels all dropped. Twelve similar wagons, with 2ft 6in sides, were built in 1902 specifically for conveyance of sand from the Wirral; P13 was the index number given later about 1910, and some other P3-type wagons were also built in 1902 for the GWR/LNWR joint lines.

Fourteen-ton, 12cu yd, ballast wagons appeared in 1899, and over 300 had been built by 1901 (P2). The underframe was like N4, 20ft over headstocks (with 10in channels) and similarly Thomas braked. Three 6ft 8in by 2ft 2½in panels fell down, the bigger capacity than before coming from an 8ft width. In many of the earliest wagons, the end and centre pillar stanchions to which the doors were pinned were 'hidden behind' the doors, so that the width of the actual drop door was the panel width. Later, the pillars were brought out to the edge of the solebar and could be seen, so that the width of the falling door was the in-clear width. This change took place in the last lots of P2, when also DC brakes replaced Thomas. The final lot in 1904 followed the then current vogue of using vacuum fittings (actually the last P2 wagon was 14009 in 1906, built as a replacement match truck for crane number 87).

The remainder of the eleven numbers issued when the diagram list was drawn up were the veritable dinosaurs and trilobites of the ballast wagon world! P1 were 20-ton bogie ballast wagons, 34ft over headstocks, 7ft 6in wide, 2ft 3in height with 6 fall doors along each side (RW p.106). The

Below: *For P14, shorten to 16ft oh, GW s/c buffers. For P18, as P15 but Morton brake (as N30) modified doorstops, steps. For P20, as P18 with extra longitudinal u/f scantlings (externally same as P18).*

Bottom: *For P21, as P19 with extra longitudinal scantlings (externally same). For P17, as P16 but 2ft 5¼in sides, 21ft 6in oh (as N29). For P23, as P17, but slotted Morton brake as P19, extra scantlings.*

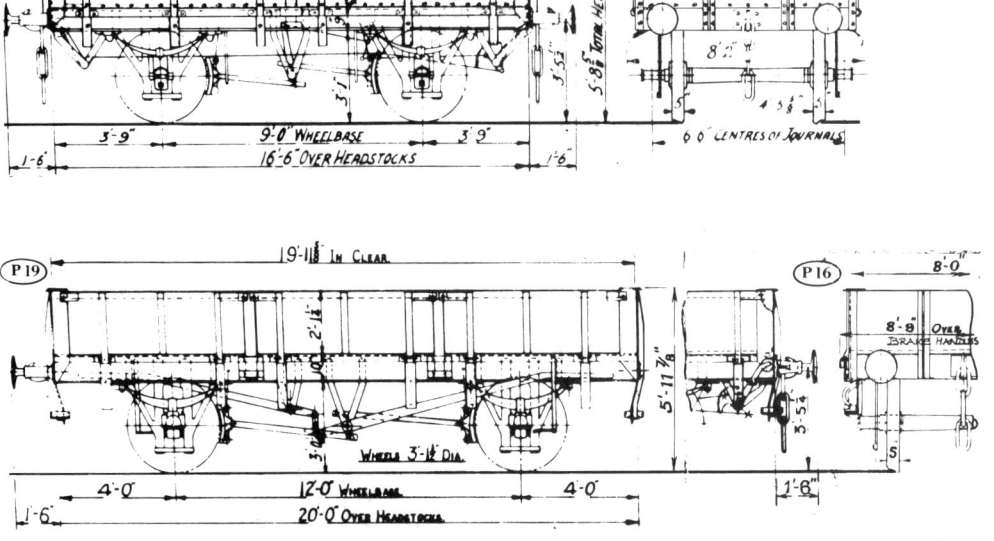

P7 hopper ballast wagon. Probably photographed when built 1893. Original 12-ton low-sided hopper. One-sided lever brake far side toothed rack. Grease boxes, covers over springs. Non-Gedge hook. Plate rail added round top and DC vacuum brakes fitted 1902–6 then given P7.

P6 pattern 20-ton hopper ballast, 10in solebars. Photographed when built 1902 with BG style livery between cast plate period and large GW. Dust covers over springs and OK boxes. Handed brakeblocks. One of the few examples of cross-cornered DC I brake (note connecting rod in front of hopper below solebar).

P2 14 ton ballast wagon. Photographed when built 1899 outside Newburn House, Swindon. Plates for number and other details. 8 by 4½in OK oilboxes. Thomas brake. Wooden doorstops. Covers over springs. All side pillars covered by doors. Final lot were built with DC II vacuum brake and exposed corner pillars, spring door stops. U/f similar to N4, but laminated sprung buffing and drawgear.

P12 20-ton ballast wagon, built 1910, photographed 1920. Only ballast wagon with external T-stanchions (do not tuck under solebar). U/f similar to P2, except lower doorstops. End lettering behind vacuum pipe (cf OPENS B).

P14 10-ton ballast wagon built 1913, photographed 1920. GW oilboxes. Only ballast wagon with GW self-contained buffers. Cast plate for Wolverhampton Division out of period.

design dates from 1890, and in addition to the 15 'narrow gauge' wagons that were built, 15 others were constructed convertibly for the broad gauge as late as 1891. The underframe was similar to that of the TOURN; the original grease boxes and lever brake on the bogie were eventually replaced by oil boxes and DC brakes. The design appears to have provided for shovelling over the ends of the wagons, an unusual feature; two of them (40514/5) were equipped with their own ploughs (some of P7 also were equipped with spreaders, such as L30). P8/9 were very small tipping wagons used by the construction department for special earthworks. P8 were 61 six cu yd side-tip wagons built in 1898 (RWA Fig 268), and P9 was one end-tip wagon built for the construction of the Fishguard Harbour line in 1898. Similarly, P10/11 were narrow-gauge quarry wagons, dating from 1890 (four wagons for Dulcot Quarry) and 1902 (three wagons for Black Rock Quarry, Tenby) respectively. All these vehicles, including the standard gauge P8/9, were dumb buffered and unsprung.

In 1907, 20-ton (15cu yd) four-wheel open ballast wagons were introduced, and were the first wagons added to the index as P12. Like P2, they used a loco coal underframe, by this time the 21ft over headstocks N2. The enlarged capacity was obtained by increasing the width and height. The wagons employed T-stanchions, so that over the door hinges the vehicles were 9ft wide. This was evidently too restrictive, since subsequent 20-ton designs reverted to an 8ft width and omitted stanchions, and the 84 wagons of P12 built by 1910 were the only ballast wagons to have external T-stanchions. The side doors were similar to P2, the extra foot length being taken up by the 6in end corner pillars which were brought out to the edge of the solebars.

Only forty 10-ton wagons had been built on two lots in 1896 and 1901 to the P3 design. The number of 10-ton wagons was increased in the years immediately preceding World War I by the construction of P14. This 1911 design had two doors each side, 7ft 3in in clear, with two end 6in pillars and, for the first time, a centre 4½in pillar. The underframe, with self-contained buffers, was like O11, and P14 were the only ballast wagons to have GWR self-contained buffers (RWA Fig 288); 300 wagons were originally ordered, but only 270 were actually built because of the war. The uncompleted third lot, 793, was finally closed out on the books in 1928.

No new ballast wagons were constructed thereafter at Swindon until 1935, and then 10-, 14-, and 20-ton wagons were built conjointly. Although eight new diagrams were issued for such wagons by the end of 1947, there were really only three body designs for all practical purposes. P15 was the new generation of 10-ton wagon, and was longer than before at 16ft 6in over headstocks, yet retained a 9ft wheelbase (cf N30) (RWA Figs 281/3). The height remained 1ft 10in like P14, so that more capacity was provided; since P14 was lower than P3, it skimped on volume even though all were nominally 8cu yd. P16 was like P2, but had two end and two centre pillars exposed; also the capping was less cumbersome. P17 bore the same relationship to P12, but was 21ft 6in over headstocks, using an N31 underframe; the width reverted to 8ft, so the height increased to 2ft 5in. The nearly 500 wagons of P15, 100 or so of P16 and 40 of P17 all had RCH fittings except that the DC brake was used. P18/19 of 1941 were the previous P15/16 equipped with Morton brakes, but had the additional feature of steps at the ends, the longer wheelbase P19 having a slotted link brake, Vol.1, p.79. It seems likely that L1288 of 1938, normally attributed to P15, may have had the Morton brake gear.

In the middle of World War II came a change of policy regarding the layout of members in the basic underframe, and two continuous longitudinal scantlings appeared on such as O37/8 and V36 (Vol.1, p.54). Thus in the P group, P18/19 were changed to become P20/21 in 1943; the 20-ton design was not altered until 1947 when it became P23, again with a slotted link Morton brake.

Many of P15/18 were used as replacements for loco sand wagons, working between Swindon and out-stations. Various old vans were also converted for this purpose in the 1930s. One wagon of P19 was subsequently used as a weight tender for Messrs Pooley & Son's weighbridge tool vans.

Although no ballast wagons were built between World War I and 1935, L1061 of 1930 was issued to cover the purchase of 60 patent screw-over tipping wagons of 15cu yd capacity from Metropolitan Cammell (RW p.100). The special feature of these wagons, for which no diagram number was issued, was that the body moved over to the chosen side of the wagon before tipping began, and without the centre of gravity rising. Tipping saved shovelling and hence manpower. Originally the wagons were used for carrying spoil to and from embankment slips (not all engineering construction vehicles were used for ballast). Many side-tip wagons were marked to Tytherington Quarry near Yate, and other names often seen on ballast wagons were Meldon, Nantyglo and Coalbrook Vale (Western Valleys).

CHAPTER 15
Q – PROVENDER WAGONS

The GWR had its own stores for provender – dried food such as hay fed to horses – at Didcot, the tall warehouse-like building on the Swindon side of the station. A description by W. H. Stanier of its workings, oat-crushing mills and so on is given in the *GWR Magazine* for October 1906, reproduced in the *Great Western Echo*, spring 1972.

The bulky nature of hay gave rise to extremely high-sided open wagons built to serve the stores. The 6ft deep bodies (nine 7in plus one 9in planks) were higher than the sides of contemporary iron MINKS! Twelve provender wagons existed and were built in two lots of six. Diagram Q1 is of the 1903 10-ton design, 18ft over headstocks, 11ft wheelbase by 8ft 6in wide. The earlier 1884 nine-ton vehicles had slightly different dimensions, including the possibility of the rather short wheelbase of 6ft; they were however amongst the first opens to be built with iron frames. The DC I arrangement on the later design is of note because of the offset V-hanger. Sadly no provender wagon photographs have been located, save for a distant view of one in a prospect of the Didcot stores, and

the close-up of an end of one shown with the old three-planker. The side doors were very much like van doors (cf V5, Y2) but were wider.

There was much more traffic in general fodder than could be carried in these six wagons, and sheeted opens were employed, for example in the movement of hay from the West Country to the temporary forage store at Newbury in the first World War I.

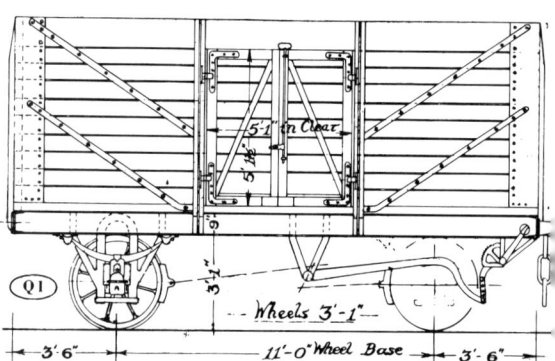

CHAPTER 16
R – MANURE WAGONS

There is specialised traffic on a railway and perhaps extra specialised traffic, and the quintessential manure wagon must come into the latter category! Six vehicles (36980-5) were built to the R1 design in 1905. In the days of horse road deliveries, when the GWR worked many thousands of horses, disposal of manure was a significant operation. However, the principal use of these wagons was at Fishguard Harbour in connection with the Irish cattle traffic. Extensive lairage was provided

there and these wagons serviced the cattle pens. The lots show that although no further manure wagons were built as such, many old OPENS were converted for the purpose, and lettered 'Return to Fishguard' or 'Return to Carmarthen'. For example, L1227 refurbished OPENS 72232 and 10014 to replace 55252 and 11482 in 1936 and L1351 replaced 25257 with 63307. Of the six R wagons only 36985 was in traffic in 1936 and eventually was used for carrying beer barrels (RW p.128).

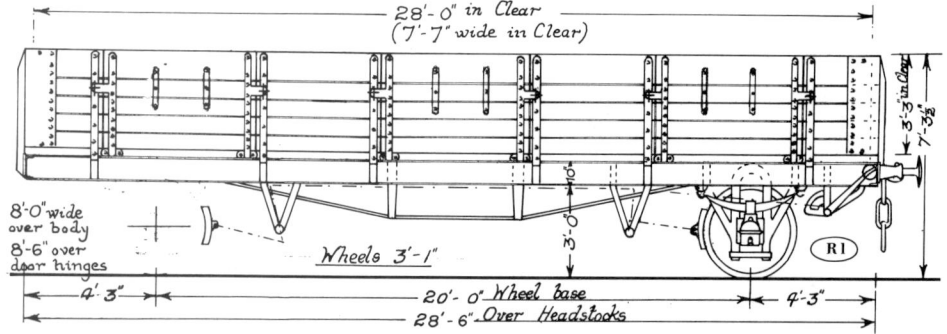

CHAPTER 17
S – FISH WAGONS
(Tadpoles and Bloaters)

In the early days fish in boxes and barrels was carried in open trucks from the West Country and at the conversion of the gauge in 1892 the special wagons used for this traffic were altered to run on standard gauge. Fish *vans* did not appear until 1909, and even then were not immediately classified into the S group, this being reserved for open wagons only. Many of the open trucks were relatively old when the S index was set up (indeed the earliest were themselves 1878 conversions of carriages) and many were condemned soon afterwards, leaving gaps in the index; the passage of time and loss of records has led to much conjecture about these low numbers in the index. The situation has been aggravated by other circumstances, viz: of the relatively few photographs available, most are of broad gauge vehicles; the original broad gauge lots do not always refer to 'fish trucks' as such, but sometimes merely to open wagons; the vehicles altered their telegraph codes subtly after the diagram index was set up (at first the A-suffix on TADPOLE meant 'with a guard's box' whether 6 wheel or bogie), but circa 1909 it became the code for all bogie fish trucks (with or or without guards' boxes); many wagons were uprated from their original tonnage, and most Swindon drawings do not help chronologically since they are from the period 1904-15 after the uprating; finally the vehicles were renumbered into the van list in 1915-16 (changing from grey to brown livery) and as not all the original vehicles had survived, the re-numbering has been mis-

understood (indeed some BG vehicles were scrapped directly without conversion to NG).

In 1905 seven S diagrams were allocated to the extant open fish wagons, which by this time did include some new standard gauge construction. Bogie, six-wheel and four-wheel trucks were represented. The 'newest, longest and heaviest' was the five-plank 12-ton 34ft 7in over headstocks vehicle then recently built on the frame of the BG Queen's Saloon, under (passenger) L1047 in 1904, given goods vehicle number 42809. Some 40ft over headstocks bogie trucks which had been built convertibly under osL473 for the broad gauge in 1889 came as S2. Originally they were eight-ton, four-plank wagons with a guard's box in the middle, numbered 11210-16, and became 42801-7 upon conversion (Vol.1, p.62) at which time some of them lost their cabins. One standard gauge 12-ton bogie fish truck with guard's box had been built in 1891 under osL620 to a three-plank design. This vehicle took the sequential number 42808, and was given S3 in the index; logically S2 and S3 should have been inverted because of age and tonnage, but it seems that their numbering order was followed. A standard-gauge three-plank bogie fish truck, without guard's box, was ordered under osL619 at the same time as S3 in 1891, and assigned 42894, but it was cancelled. S3 later lost its cabin during World War I, which produced a plain three-plank wagon (RC, p.38) which is presumably what osL619 would have looked like. The bogies of S1/2/3 were the Dean corner hung design,

S1 TADPOLE A (post-1909 code), built 1904 on broad gauge u/f of old Queen's saloon, Dean corner hung bogies. Photographed 1920. Curious colours on repairs, grey and brown. Lettering (still large) seems white. 'V' and wheelbase etc information on solebars.

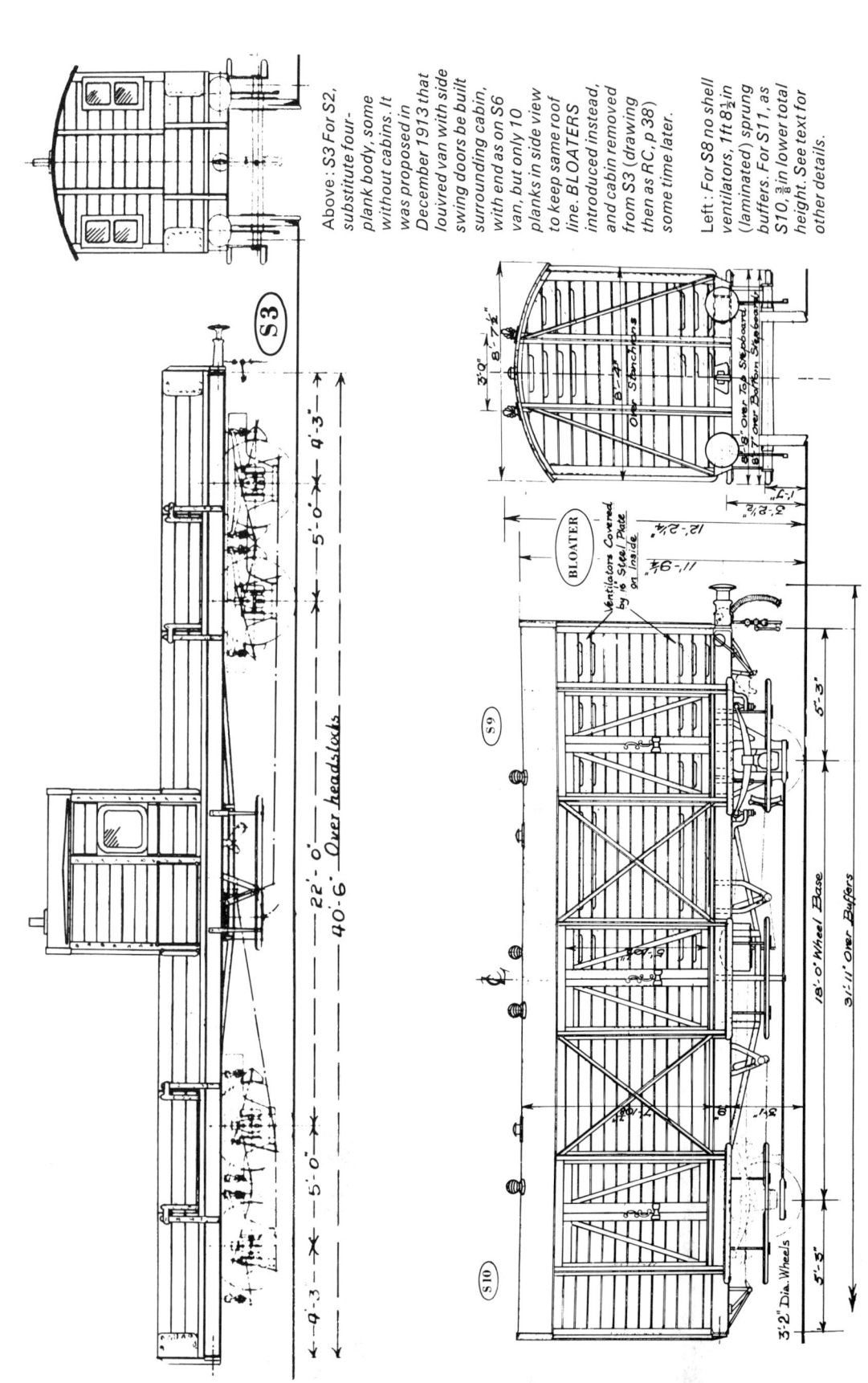

Above: S3 For S2, substitute four-plank body, some without cabins. It was proposed in December 1913 that louvred van with side swing doors be built surrounding cabin, with end as on S6 van, but only 10 planks in side view to keep same roof line. BLOATERS introduced instead, and cabin removed from S3 (drawing then as RC, p 38) some time later.

Left: For S8 no shell ventilators, 1ft 8½in (laminated) sprung buffers. For S11, as S10, ⅜in lower total height. See text for other details.

S3

9'-3" — 5'-0" — 4'-3"

22'-0"

40'-6" Over headstocks

9'-3" — 5'-0"

BLOATER

S9

8'-7½"

3'-0"

8'-3" Over Stanchions

8'-3" Over Top Stepboard

8'-7" Over Button Stepboard

1'-7"

3'-2½"

12'-2½"

11'-9¼"

Ventilators Covered by 18 Steel Plate on Inside

5'-3"

18'-0" Wheel Base

31'-11" Over Buffers

S10

5'-5"

3'-2" Dia. Wheels

S2 having 4ft 10in or 5ft wheelbase, S3 5ft and S1 6ft 4in Dean bogies.

Six-wheel trucks (all of which had originally been broad gauge vehicles) made up diagrams S4/5. OsL181/2 were a rag-bag collection of vehicles built on old broad gauge carriage frames including those of 'iron' coaches. Their dimensions varied markedly and some had guard's centre boxes, none of which fitments survived conversion to standard gauge. OsL420 were 36 new broad gauge convertible wagons built in 1887-8 (cf 42795 in RC in original six-ton form with spoked wheels). After conversion some of the six-wheelers were uprated to eight tons (with seven-leaf, instead of six-leaf, heavier springs) which at the time of indexing were given S4 (Vol.1, p.72), the remaining six tonners being given S5. There was a one-off six ton ducketted brake van version (with steel panels on the cabin) of the last six-wheel wagon design, built under the next lot osL421 in 1888 (broad gauge 11209, standard gauge goods 42800 and van 2045), which was lumped in with the rest of the six tonners in S5. It has been suggested that this wagon was given the S8 diagram, which seems to stem from a misreading of a stencilled '88' on the drawing; S8 is logically wrong anyway. Moreover, the 6 wheeler with guard's cabin was not condemned until 1926, and there is no dispute that the eventual holders of the S8 diagram (BLOATERS) were introduced in 1916 (see later). The most obvious difference between the various Swindon diagrams for six-wheel fish trucks is the length of the springing. The six ton brake van drawing with 8in by 4in journals (osL421) has 4ft 6in springs, as does the 8-ton drawing with 8 by 4in journals which is definitely marked S4. The six-ton drawing with 8 by 3½in journals however is shown with 6ft 6in springs,

the only other obvious difference from the eight ton drawing being the height of the underside of the solebar from rail level, 3ft 2in on the six ton wagon instead of 3ft 4in. That these drawings do not give the whole story is readily seen in photograph in RWA Fig 4 of the eight ton rebuild of 42790 (formerly broad gauge 11199, later van 2035), and the aforementioned S4 truck in Vol.1, which have 6ft 6in springs. It may be that the last few of osL420 had short springs; the earlier wagons built on carriage underframes most probably had long springing.

Many four-wheel broad-gauge carriage underframes were also used to make fish trucks, and these vehicles came at the end of the list as S6/7. Fifty four-wheel six-ton poultry and fish trucks were assembled in 1891-2 under osL608 all with varying dimensions (S6). The four six-ton vehicles (42615-18) remaining from the six of osL343 (11162-7), which had been made up for the broad gauge in 1885 from old carriage frames, became S7 in the index; after 42616 had been condemned in 1910, the survivors received brown vehicle numbers 2071-3. As shown in the frontispiece photograph of this book, these four-wheel vehicles had bars along the sides with planked doors and ends (drawing RC, p.39). In Vol. 1, it was erroneously suggested that osL343 were six-wheelers and thus in S5; the two vehicles 42551 (ex 999) and 42552 (ex 1000) assigned to S7 in Vol. 1 are in error since they were condemned as the index was being drawn up. Note that the serial numbers of S5 osL181 in Vol. 1 Chapter 3 should start at 42553.

Some low and high six-wheel SIPHONS built at the turn of the century were used as fish vans until 1908, but then fifteen 10 ton V12 type vans were constructed as part of L625, with extra slats

S6 fish van built 1912 as 85843. Photographed 1929 in brown livery. Body based on V7, but sliding doors. Vacuum and steam connections. DC III white stripe. External dimensions on solebar in block alphabet. Label clip on body. Lamp irons.

Left: *S8 BLOATER A built
and photographed 1919.
Brown vehicle, but seems
white lettering. 'Westinghouse
Brake' under far right door
and on end; star and W under
centre door. Sliding doors,
gas lit. Side lamp irons.
Underframe similar to T8/V11.*

Left: *S5 TADPOLE A
(pre-1909 code), built 1888
as BG 11209, later van 2045.
Short springs, brake
compensating rodding in
front of axleboxes.
(Courtesy Locomotive
Publishing Co)*

Far right: *S2 built as V13 in
1909 and formerly marked to
South Dock Swansea.
Photographed with 'new
writing' and GWR totem in
1944, when post 1943 FISH
code appears. Brown
vehicle. Label clip on body,
far right steam connection.*

between the side planks and in the plain ends, and
were reserved for fish traffic with the diagram V13
(82876-90; later van 2074-88). They were followed
in 1912 by 25 fish versions of the 21ft over headstocks V7 design with slats along the tops and
bottoms of both sides and ends (no shutters or
bonnets). These vehicles were the first vans to go
into the S index, taking over S6. All but two of the
odd collection of fish and poultry trucks had been
withdrawn by 1912, and although unusual, it
seems that the new fish vans (85831-55) ousted the
survivors from the S6 diagram. Perhaps it was

thought that the old trucks would soon be scrapped
(otherwise S8 would have been issued then); on
the other hand they were given van numbers
(2046/7), and not scrapped in 1915-16 as could have
happened if they had been 'forgotten' in 1912.
In the event they lasted until 1921/6 respectively.

In 1913 a proposal was put forward to add a
van body with slatted sides like S6 around the
guard's box of TADPOLES A, but instead the
new equivalent BLOATERS were introduced in
1916. The bogie van would have been 12-ton
capacity with an estimated tare of 15 tons, whereas

Left: *S13 INSIXFISH. Built
1948, photographed 1965
(INSUL-X-FISH code).*
(D. J. Hyde)

Right: *Side, end and
plan views of S13.*

the BLOATERS were four-wheel 10-tonners with tares about 12 tons. They, like S6, were designs adapted from merchandise vans, in this case the 18ft wheelbase 28ft 6in over headstocks V11 but with three doors per side to give quick access to the perishable fish; again, like S6, sliding doors were used. These gas-lit BLOATERS were built between 1916 and 1928 under carriage lots and directly were given van numbers, since the brown vehicle classification had appeared during World War I, Vol.1, p.86. Four diagrams (S8-11) were issued and covered minor variations in design: all except the first lot of S8 had self-contained buffers; S11 were built by Metropolitan Cammell and were marginally lower than the others; S8/9 were the only designs to have ventilated sides but were slightly different, S8 having no roof ventilators; S11 had a 'Decolite' fireproof (instead of an asphalt) floor, ribs on the roof and no end louvres. Floor drainage traps were different on the various designs; 39 were dual-braked, giving rise to the code BLOATER A.

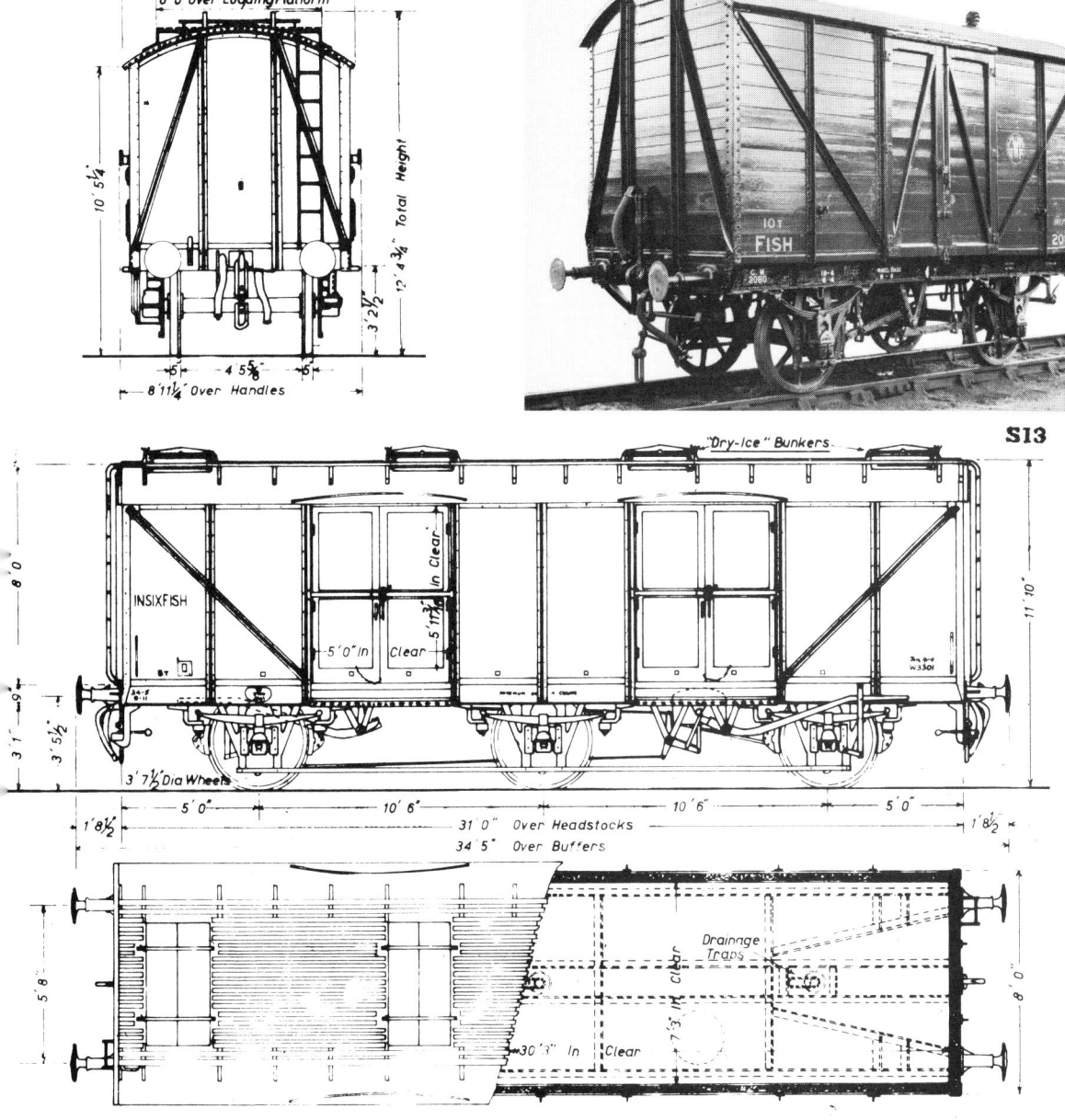

S13

The fleet of vans was augmented by 20 further S6-type vans in 1925 which, because of the absence of side vents and the use of self-contained buffers were given the separate diagram S12.

The scrapping rate of TADPOLES increased rapidly after the introduction of BLOATERS; of interest the 4-wheel 2057 was made a truck (166) for the Swindon munitions factory in 1918. Nevertheless all the six-wheelers of osL420/1 and all the bogie wagons (except 42802, withdrawn in 1913) were still running when the end came in 1925-6. Many were then condemned from the Foss Cross storage sidings, and the bogie trucks 2001/4/5/7/8 were the last to go, on Christmas Day 1926.

In 1927 the V13 fish vans were transferred into the S index. Why they were given S2 is a mystery, since S1/2/3 were all vacated simultaneously as just mentioned, and S4/5/7 were already free. It has been suggested that the V13 vans were the original holders of the S2 diagram on the supposition that the S index was set up circa 1910 later than the V series of 1905. The evidence for a 1910

date is not clear, and it seems odd that 10 ton 4-wheel vans should be placed ahead of the 12 ton bogie brake (S3) in a first listing. Moreover the transfer of vans out of the V index in a 1920s reshuffle has parallel with the Y index banana vans.

Not until just after nationalisation were new fish vans introduced. The INSIXFISH design was a development of the insulated Palethorpe's sausage vans of 1936, where insulation was provided by 3in layers of 'Onazote' in the cavity body construction. Removable slatted wood mats allowed free air circulation round the fish boxes. The walls (of resin bonded plywood) were treated with shellac and varnish on the inside to give a clean non-absorbent finish. The outside was painted cream, with black underframe, lettering, door fastenings and handles. Four dry ice bunkers were provided in each van, with a capacity sufficient for the storage of fresh fish for up to forty-eight hours, which was necessary on weekend loadings. The vans were used on the Penzance – London, Milford Haven – Birmingham and Milford Haven – London routes.

CHAPTER 18
T – PERMANENT WAY AND SLEEPER WAGONS

Work on the permanent way could be of three main types: relaying or complete renewal of the track (sometimes with secondhand rails); re-sleepering and re-railing. For example, in 1929, 227 miles were relaid with new materials, 150 were relaid with secondhand rails, 115 received new sleepers and 27 new rails. Almost all crossing work was constructed wholly of new materials, although secondhand rails were sometimes used for siding points. Three distinct varieties of vehicle were found in the T group for this work, wagons for chaired sleepers, for unchaired sleepers and for the long sleepers of switches and crossings (vehicles of this last type were also used by the signal department for signal posts); rail wagons were of course in the J group, and ballast in P.

T1/12/13 were 18-ton well wagons for chaired sleepers. Bearing a certain similarity to the CROCODILES, T1 dated from 1894, and 101 vehicles were built by 1900. Tall side stanchions allowed the load to be carried quite high. Eighteen new vehicles were built in 1938 to diagram T12. Essentially the design was like T1, except that the wheelbase was lengthened by 6in to 25ft, the overall length remaining as before at 32ft; 1ft 8½in large self-contained buffers were used since

vacuum fittings were provided. The side stanchions were split in two for ease of removal, and pockets in between the stanchion bases were provided to store chains and shackles. T12 was one of the last new designs to use DC brakes; a further six wagons built in 1942-4 with an independent lever brake were given T13. The LORIOTS W of 1931 and 1944 had similar features to T12/13.

The wagons for carrying plain sleepers were like open wagons with no doors and were characterised by round ends which gave 2ft 4in in clear and 3ft 6in at the ends. The early 10-ton wagons (T2/6) were 19ft over headstocks, the later 14-ton wagons (T8/11) being 28ft 6in long. Fifty-one vehicles were built to the T2 design between 1893 and 1901, and four more appeared in 1911-13 to the T6 diagram. These latter replaced some wooden solebar six-ton trucks that had been built to a T1-like design in 1877 under osL170. Incidentally, these old wagons were four-plankers and predated the introduction of merchandise four-plankers by ten years. T6 was 8ft wide, rather than the 7ft 6in of T1, and omitted the curious feature of four end stanchions. The underframe of T6 also lengthened the wheelbase from 11ft to 12ft, used self-contained buffers and replaced the simple

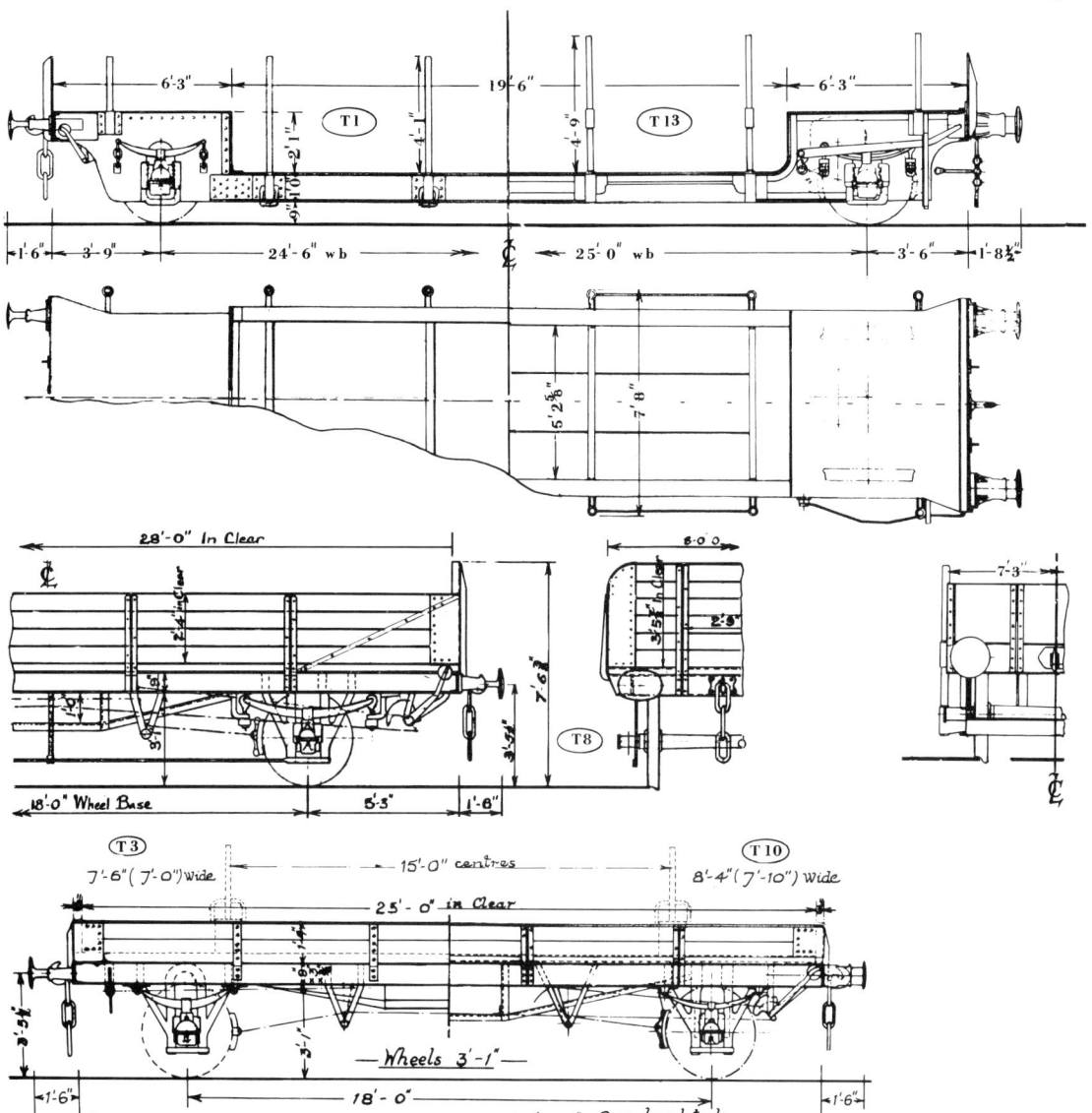

one-sided lever brake by cross-cornered DC brakes. Twenty-seven T8 wagons appeared during World War I. Essentially stretched-out T6, with 18ft wheelbase, their interesting features lay in the underframe, which was copied by the BLOATERS S8, with long underslung springs. The two vehicles of T11 in 1921 have an S9 type underframe with 16in self-contained buffers and drawgear, Gedge hooks, and instanter couplings; the buffers however were 1ft 6in long, not 1ft 8½in. None of these wagons were built with side doors. Most of T8/11 were given the same running numbers in the 14xxx series of the condemned osL170 wagons that they replaced. However on L875 (T11) 14679 replaced 14460, not as stated in Vol. 1 Chapter 3.

The third variety were all two-plank opens. T5

Top: For T12, as T13 with DC II brake each end. End view is to the right below the plan view.

Centre: For T11, substitute GW s/c buffers. For T6, alter to 19ft oh, 12ft wb, GW s/c buffers.

Bottom: For T4, widen T3 to 8ft (7ft 6in in clear) with DC III brake, modified trussing. For T7, like T4 but widened to T10. For J7, shorten wb to 16ft, 6ft 3in in clear between bolsters.

Above : *T1 chaired sleeper wagon built 1894 with no plate ends. Photographed 1898. Original one-sided lever brake (later DC II), toothed rack. OK 10 by 5in oilboxes, 3ft springs, D shackles. Buffing and drawgear springs protrude through side plates.*

Left : *T2 sleeper wagon as built 1897. Tare painted, load 10 ton. Far plate says 'GWR. When empty return to . . . (firm ?) . . . Newport'. Private owner registration on plate (including load) ? GWR 8 by $3\frac{1}{4}$in oil boxes. Note V-hanger arrangement, inside u/f is vertical rod.*

were two 10-ton wagons built in 1907 for the signal department based on an earlier, narrower design built under osL222 in 1880 (RWA Fig 296). They had a length of 20ft, a 12ft wheelbase and the unusually large width of 8ft 4in (cf P12 about this time); DC III brakes were employed. The rest of the two-plankers in this group were all 14-ton wagons, 25ft 6in over headstocks and 18ft wheelbase, for switches, crossings and signal posts, differences arising in width and fittings. The oldest date from 1897: T3 were 7ft 6in wide and had the Thomas brake when built, RWA Fig 295. An adaptation of their body provided the basis for the 16ft wheelbase MACAW A of 1902. T4 of 1908 was 8ft wide and was built with DC III brakes. T7, made from 1910 until the war, was wider again at 8ft 3in. This width (7ft 10in in clear) followed on T9/10 into the 1920s, differences here relating to angle trussing instead of bar trussing, and additionally on T10 the use of GWR self-contained buffers.

The sleeper wagons usually had legends saying 'Return to Hayes Creosoting Yard' or 'Sleeper Traffic. Return to Hayes', and those two-plankers used by the signal department had 'Empty to Reading'.

T8 sleeper wagon. Built and photographed 1914. No doors. Cross-cornered DC III brake (handles not painted white); white stripe not central. U/f like BLOATER. Square shank buffers.

T3 signalpost (or sleeper) wagon. Built and photographed 1896. Thomas brake, 2 blocks this side with rigid connection at shoes. Flat side stanchions, OK oilboxes.

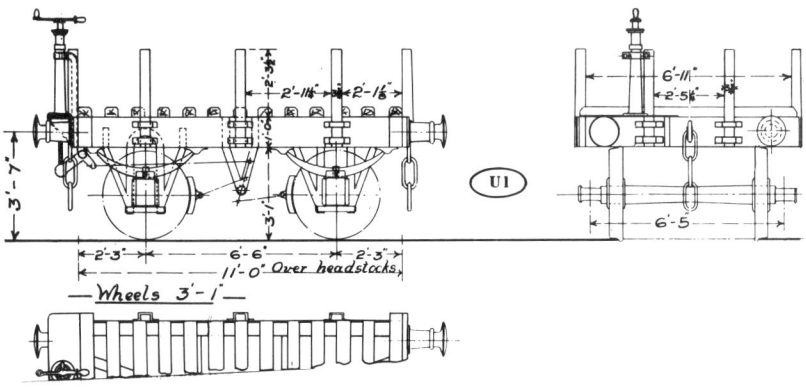

CHAPTER 19
U – STONE WAGONS

Two rather obscure wagons numbered 31001/2 comprised this group. They were ex-Cornwall Mineral Railway Nos. 1/2, built by the Swansea Wagon Co in 1874 (not 1870 as suggested in Vol.1). The 6ft 6in wheelbase, 11ft over headstocks size demonstrates a good ability to negotiate sharp curves in West Country quarries. The 1ft deep wooden underframes were quite massive for the 15 ton load. The GWR running numbers were assigned on amalgamation of the CMR into the GWR. It is not clear why they were not put into the P group instead of being given their own index. Both wagons were condemned on 9 June 1917 at St Blazey.

CHAPTER 20
V – COVERED GOODS (& GRAIN) WAGONS
(Minks, Grano)
(including general remarks relevant to fish, fruit
and other vans).

Much traffic in the very early days used to be carried in open wagons, covered by tarpaulins if necessary, and as late as 1926 some trial bulk grain open wagons were built (O25). Covered goods vans with wooden bodies first appeared in the 1860s, the underframe being wood (later bulb and channel iron underframes), the brake gear at first simple wooden-block one-sided 'arched' lever. Construction of these outside framed wooden vans ceased when in 1886 the GWR embarked upon a series of metal bodied and roofed 9 ton 16ft over headstocks vans, the famous iron MINKS (V6). Later vans were 10 tons in capacity, becoming 7ft 4in width in clear instead of 7ft 1in. Over 4000 iron vans were built by 1901 (including some 18ft long vehicles on, for example, L9 in 1893). Another diagram (V15) relates to 300 iron vans purchased from Messrs Spiller & Baker in 1913. They had been built privately for that company in 1905 to carry flour from their granary in Cardiff as they maintained that the GWR was incapable of supplying sufficient wagons for the traffic. Iron vans were used by other railway companies, and many came into the GW fleet from the Welsh railways after the Grouping, (as indeed older wooden framed vans had come to the GW from the South Wales Railway in the 19th century). Brake gear progressed from one-sided lever to DC I and III after repairs, and even vacuum fittings on a few. Late repairs put wooden doors on some iron MINKS (BGW p.72).

The swansong of iron vans came with the bogie MINKS F in 1902. These 30-ton wagons (eventually given V1) had 8ft inside height and were 36ft over headstocks (cf J5, O1); a single end DC II system with brakes on only one bogie was employed. They came at a time when indecision reigned about large capacity wagons, and certainly about the use of iron for merchandise wagons. Of the five ordered under L401 and allotted 69996-70, only 69996/7 were constructed and they took two years to build! The sequential L402 was for five bogie open goods wagons (assigned 42903-7), the design of which is not known, but they were cancelled completely. The potential use of bogie vans in fitted freights, however, led to six more MINKS F being built in 1911 with instanter couplings, vacuum equipment and centrally positioned

Above: *VI MINK F as built 1904 with single ended DC II non-vacuum brake. Round shank oval buffers. 5ft 6in plateframe bogie in original form (no external ribs) 'Square' boxes. Flat strip trussing, adjustable kingposts. Iron roof with exterior bands. Lettering is experimental (no serifs at all on G, cf painting trials for large letters).*

Left: *Outside framed wooden bodied goods van. Photographed in retirement as a sand van (cf coach bolts for interior partitions alongside door posts and vertical lever below solebar under door). Built in 1885 under L339, the last full lot built with bulb-section solebars (first lots in 1879) but has atypical stamped bearing spring shoes (also later oil boxes); usually on bulb solebars, the W-irons and cast spring shoes were attached to hard wood blocks mounted inside the frame. Early similar vans built with wooden solebars (carved letters and details, as Vol.1, p.53), one-sided lever brake (single wooden block) grease boxes and early pattern buffer etc.* (R. Metcalf collection)

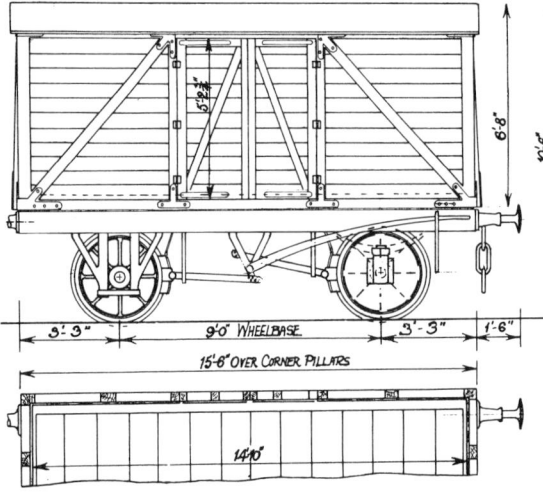

Right: *Circa 1900. Grease boxes, one-sided lever brake. Pillar support inside V-hanger. 8mm/ft scale.*

15ft 6in covered goods van. Wooden u/f on earliest, bulb u/f on L199 December 1879 (wagon no.1939), and subsequent lots, up to L358 (no.27655) which was the only outside framed wooden goods van to have channel u/f. Diagonals of body go other way on some.

DC ratchet; three took the remaining numbers originally assigned, and the others were given 79598-600, Vol.1, p.51. Bogies with top stiffening ribs were employed.

Construction of wooden merchandise vans was nevertheless decided upon from the beginning of the century, and thereafter two body styles for doors and ends characterise most vans, both in ordinary and long wheelbase. The design of door used until the mid-1920s was horizontally planked with outside wooden framing and diagonals (cf Y2); each door swung on three butt hinges attached to the adjoining T-stanchions. The van body designed for the 17ft 6in RCH underframe in 1927 (V21) had vertically planked doors (five planks per door), swinging on two strap hinges, a design that was perpetuated into nationalisation. The first vans had no ventilators, and flat X-bracing behind the end T-stanchions. After about 1906, louvres and wooden shutters were incorpor-

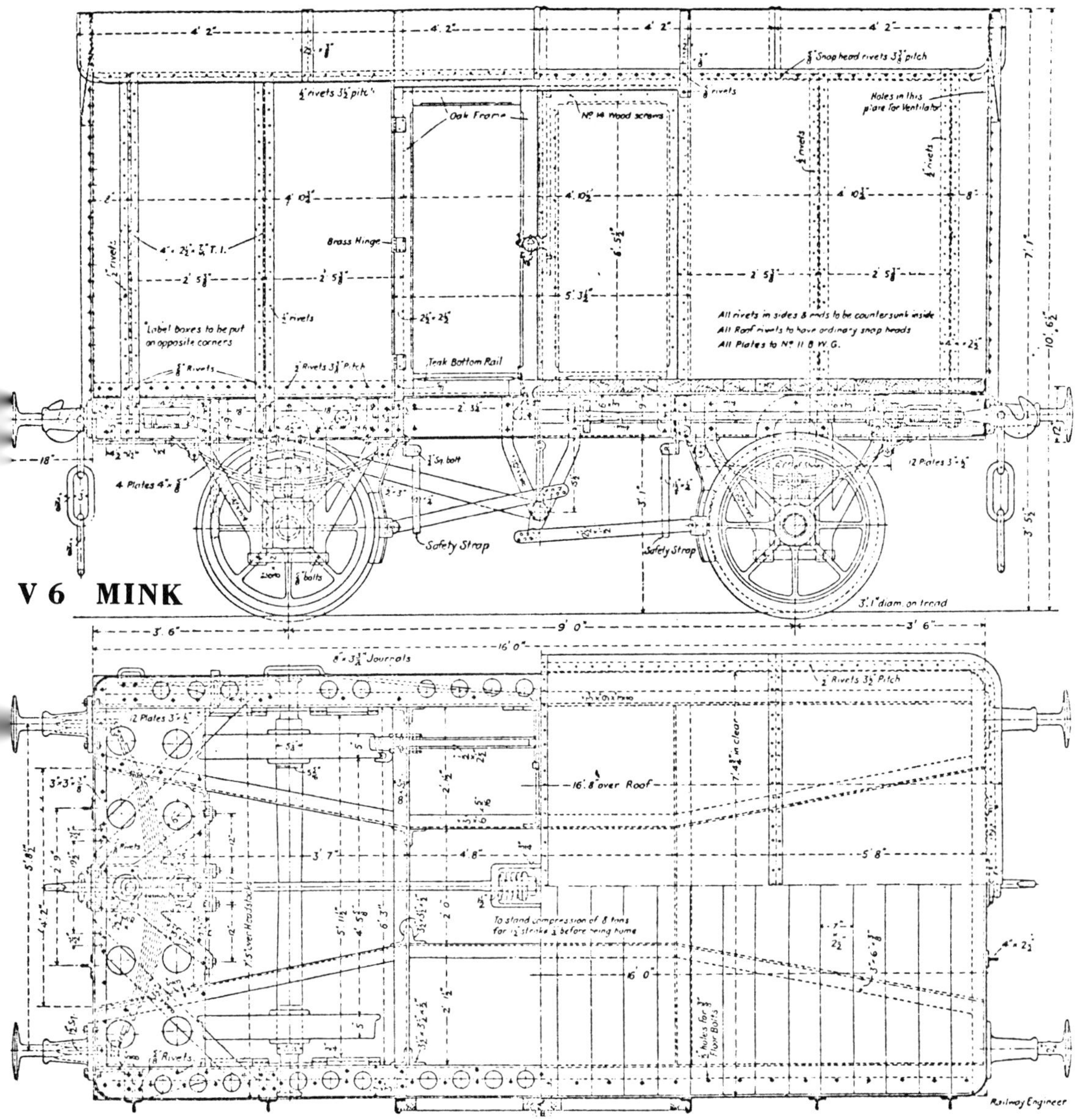

V 6 MINK

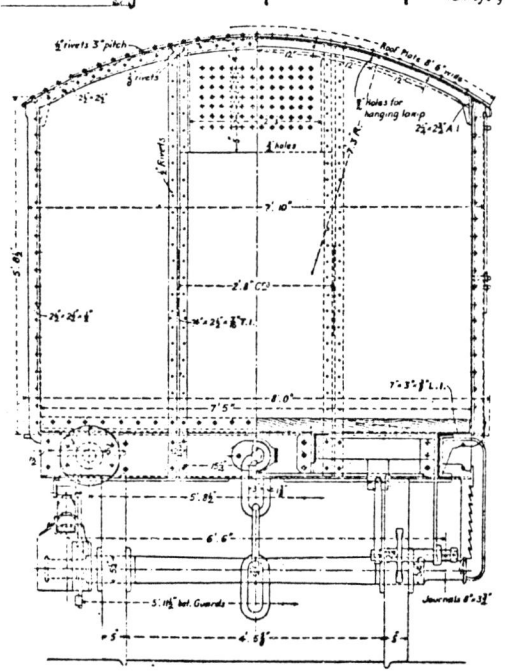

ated on new construction, the louvres taking the top halves of each outside end panel following the pattern of the V8 banana vans. V14 of 1912 had end bonnet ventilators, and for a time these too had shutters and handles, a practice discontinued in the early twenties. All these details apply in general to fish and fruit vans, with specific variations especially in regard to ventilators.

V5 of 1902 was the successor to the iron MINK, built on a 16ft underframe with DC I brakes. The inside height of these 325 vans was 7ft (some 6in more than V6/15), but in 1905 wider and still higher vans appeared with 7ft 6½in and 8ft 0½in in clear (cf Y1). Both these heights were built under one lot (476) of 200 vehicles, and put into one diagram V4; the 178 taller wagons were rated at 10 tons, the smaller at 9 tons. Almost immediately the 10-ton vans were equipped with vacuum

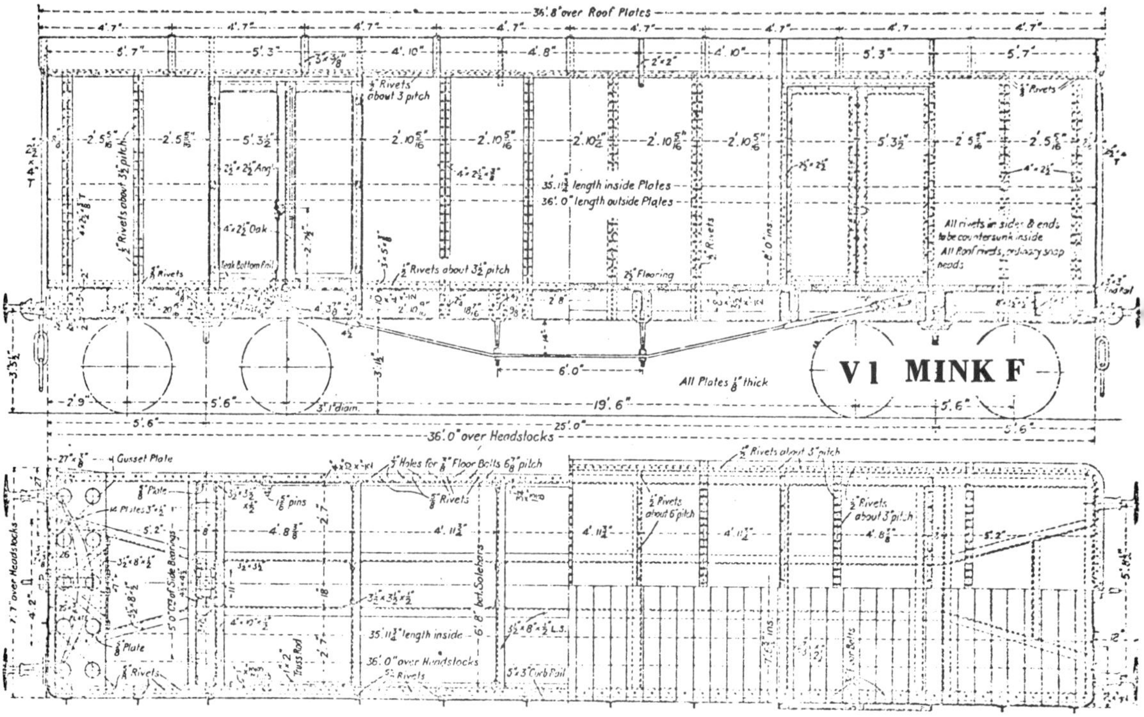

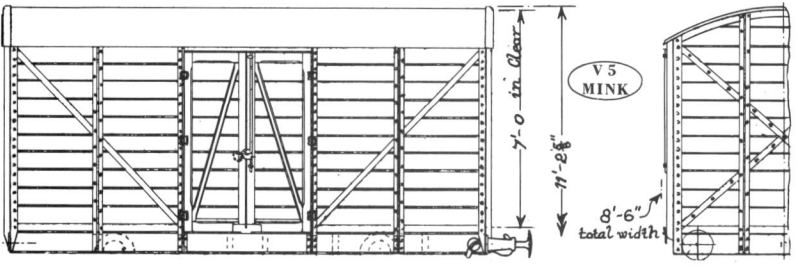

V 1 MINK F

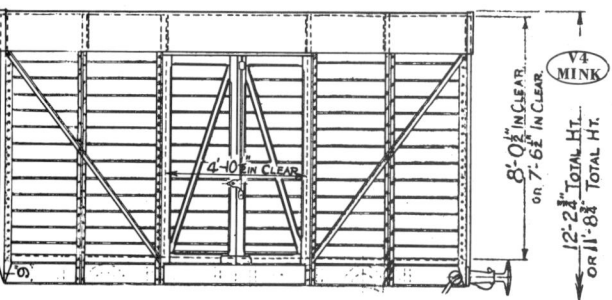

V 5 MINK

Left : Underframe as on O4.

V 4 MINK

Left : Underframe as on O4, (taller) end as V5. For V12, 7ft 6½ in height, vacuum u/f as O15 with laminated buffing gear, end as V7. For V13 (S2), as V12 with two rows of side slats near floor level. For Y5, as Y12 with plain end as V5. For V14, as V12 with GW s/c buffers, and bonnet end vents, as V22. For V16, as V14, but non-vacuum.

V5 MINK A built 1903 photographed 1965. Low height wooden MINK (cf Y2). DC I brake etc.

(D. J. Hyde)

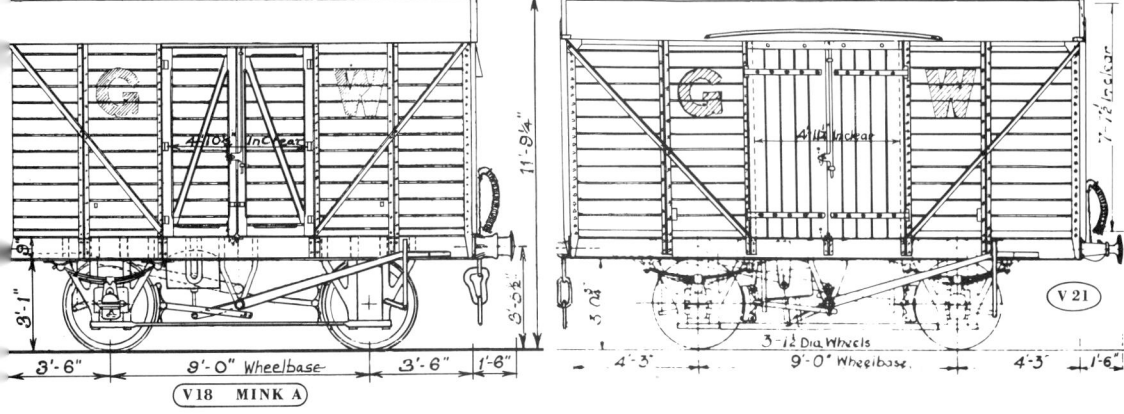

V18 MINK A

V 21

fittings for the new express freights, although the diagram shows the non-vacuum DC I underframe. When built the vans were provided with a sliding door in the roof for crane loading, but the feature was discontinued; in later years, shuttered ventilators were added to some, thus altering their code from MINK to MINK A. Two extra 7ft 6½in vans were built in 1905-6 for the Loco, Carriage and Running Department at Swindon as 'vacuum cleaner vans'. This intermediate height was made standard on all successive vans into nationalisation (actually 7ft 7½in with later bonnet vents). Then, the side-views of V12/13/14/16/17/18 and related vehicles (S2, X6, Y4/5/12) were essentially the same. The end-view of V12/13 had end louvres, but V14 and all following diagrams had bonnet vents. As regards underframes, V12/13 had DC III brakes (some vacuum with instanter couplings) with transverse sprung buffers; from 1912, self-contained buffers appear on V14/16/17, Vol.1, p.47. Some early vans in V14 from L708 had the new bonnet vents but the old transverse sprung buffers. The later lots, after 1921, of V14/16 built the vans at 12 tons, 9in by 4½in journals replacing 8in by 4in, and this capacity remained thereafter. Some of these lots were built by outside contractors (such as L983 by the Gloucester RCW). Morton brakes appeared immediately after the grouping on the 1488 vans of V18, built in 1923-5, but V14 also remained in construction until 1927, by which time some 6,000 vacuum and 3,850 non-vacuum DC III vans of all 16ft diagrams had appeared. Some of the vacuum-fitted L919 of V18 had iron roofs, and the last of these (104500) was built under the separate L954 in 1925-6 with extra planking over the external framing to give a flush-sided van. The diagram V19 was given to this special van, which was marked 'ventilated'.

The adoption of the RCH 17ft 6in underframe for the 12-ton van in 1927 coincided with a change

Above right: End as V22. For V33, as V21 non-vacuum no tiebar between axleguards. For V23, substitute 'straight down' side stanchions (not tuck under), and 'straight down' end corner angles over buffer beams as Y9. For V24 as V23 non-vacuum. For V26, as V23 externally. For V34, as V24 with extra longitudinal u/f scantlings (externally same). For V36, as V26 with plywood panels (cf G43), and extra longitudinal u/f scantlings. For V37/8 as V36 non-vacuum, no tiebars.

in door design to vertical planking, and also to a slightly wider door opening. The first diagrams V21/33 (out of sequence) retained a 9ft wheelbase with Morton brake (cf O24/6/9/36), as did the related design Y7, but after 1932 a 10ft wheelbase was used; that change on vans coincided also with 'straight down' stanchions (see O31/2). Then V23 (vac)/24/26(vac)/34 are essentially the same in overall view; the MOGOS G31 were similar, and the shortened bodies of shock absorbing vans (V27/8) were also built to the same design (for shock absorbing van drawing, see BGW p.74; Vol.1, p.91). V26 introduced the PARTO van in 1933, with internal partitions to protect loads from damage, Vol.1, p.19; among other commodities, biscuit traffic from Messrs Huntley & Palmers in Reading became a common load for these wagons. Some of these and other fitted vans were equipped with screw couplings instead of instanter. V34 had the short drawgear with no links and pins common during World War II; L1431 of 1942-4 was marked as having sliding doors. Plywood panels replaced exterior planking on L1467 in 1944 which gave rise to diagram V36, where also a 'hidden' alteration took place, in that extra longitudinals appear in the underframe (cf O37, P20; Vol.1, p.54). V36 was vacuum fitted, using instanter couplings; two running numbers were 65183/346. Plywood panels were also used on the

Left top: V19 MINK A photographed when built 1926. 'One-off' special V18 type van with flush boarding. Iron roof (bands). Morton and vacuum brakes, pinned brake rack. Door catches to hold doors open.

Right top: For V3, substitute plain ends. For V2, as V3 non-vacuum with 10 by 5in OK journals, sliding doors. For V8 (Y6), substitute Mansell wheels, T-hangers clasp shoes, central end louvres (plus louvres in central side panels on 79144 and J-hangers).

Left lower: V28 shock absorbing van. Introduced 1938, photographed 1940. Special buffing gear and side springs for sliding body. Release cord star on guard strip over side springs. Strip absent on V27, where star is painted on V-hanger (Vol.1, p.91), separate numbering on doors later omitted.

3-link coupling unfitted V37/8 which were ordered in 1946 under GWR L1524/5, but completed under BR L2083. GW serial numbers assigned were 146019-468, but only 146019-94 were built before nationalisation; BR numbers then replaced the GW assigned numbers.

Larger vans were built concurrently with the 'standard' wooden bodied goods vans. Nearly all these longer vehicles came from two groups, viz: 21ft over headstocks, 12ft wheelbase with two doors per side (MINKS B/C), amongst which some 22ft 6in oh, 12ft 6in wb fruit vans should be included; and 28ft 6in over headstocks with 20ft or 18ft wheelbase having two doors each side (MINK D) and three doors on fish and fruit vans. Most of the merchandise vans were built before World War I, so old doors and X-bracing were typical; the earliest had no end vents, then came louvres and shutters, and finally bonnet vents after 1912. Some louvres were replaced eventually by bonnets. The arrangement and number of ventilators varied with the traffic, particularly on fish and fruit vans (qv). Relatively few 'modern' vertically planked door equivalents of the wagons were built.

As regards merchandise goods vans, the first

V2 MINK B with sliding doors. Built 1903 with DC II brake, photographed with post-grouping 16in lettering. The first 21ft oh van design on the GWR. Rated at 18 tons, hence 10 by 5in OK oilboxes; 8ft clear inside height (cf V1/3/4). Later derivatives reduced tonnage and volume. Characteristic X-bracing on ends, and outside framed doors, but runners and stop blocks at middle and ends. All six vehicles (47722-7) running in 1936, but two scrapped by 1946, and only two left in 1953.

(D. M. Lee Collection)

V7 MINK C built and photographed 1907. Standard 21ft over headstocks design. DC III brakes wagon blocks. No door catches to hold doors open as built.

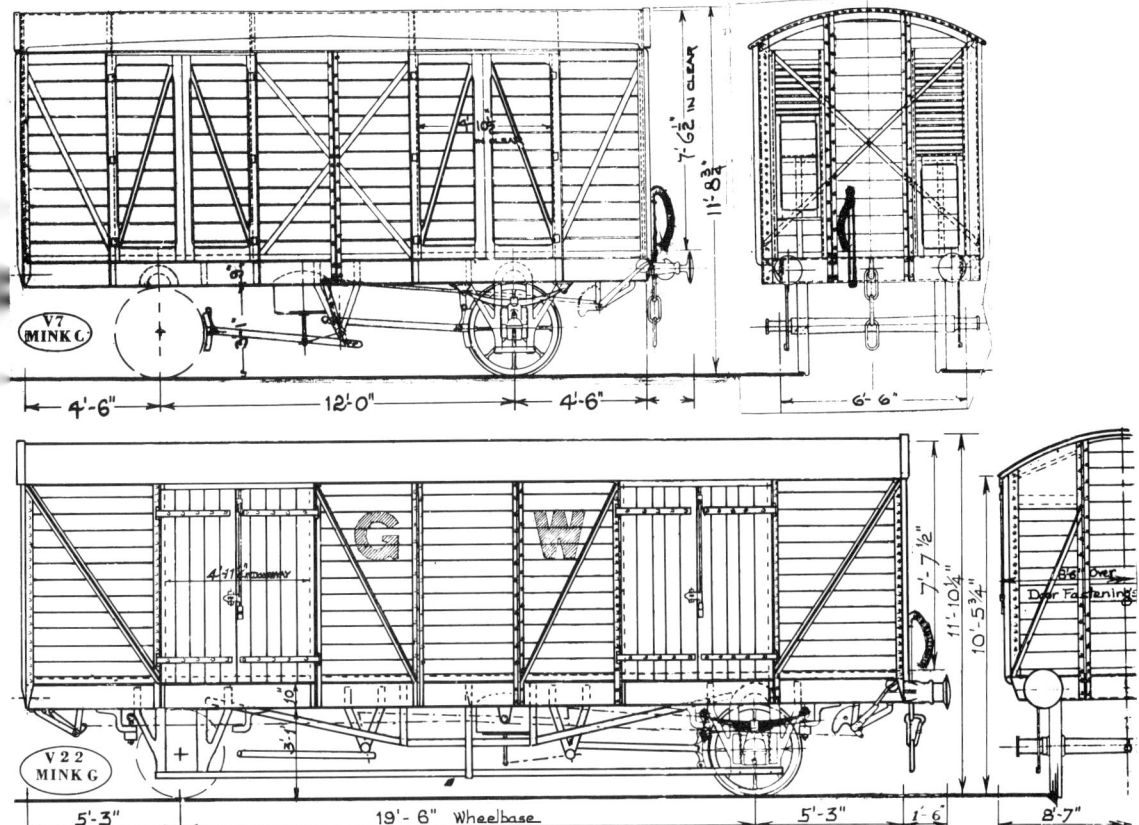

21ft design were the six V2 MINKS B built in 1903. At 8ft inside height, the same as V1, they were 1ft taller than the contemporary V5 and 1ft 8in taller than the iron MINKS. They were equipped with 10in by 5in journals, and rated at the comparatively high figure of 18 tons. An unusual feature of the design was the sliding doors, with external runners the length of the van.

The next 21ft vehicles were V3, built on the preceding lot to V4, with two similar inside heights being constructed, the vehicles having more common hinged doors and being rated at only 10 tons. Many were vacuum converted almost immediately in 1906. V7 was the same design, with the addition of end louvres and shutters (MINK C), but they were built as such with the vacuum brake (drawings show clasp shoes on early ones, akin to the arrangements on early MICAS and FRUITS, but photographs indicate ordinary wagon shoes, with tie bars generally between the axleguards). The couplings were 3-link for the fitted vans of V3/7 but some were given screw connections. The expense of screw couplings led to the introduction of the instanter coupling in 1908. Another 21ft van, V8, strictly preceded V7, being built in the previous year 1905, and was the

first wooden van of any sort to be fitted with shuttered louvre ventilators. V8 was an experimental 'special banana van', and there was an abundance of vents. Both centre panels between the doors had shuttered louvres and double louvres were fitted on the ends. In the fashion of the times, V8 had underslung 3ft 7½in wheels and 2ft buffers for passenger traffic. Twenty more were built the following year, without the side vents and with only a single central end louvre; they also had T-suspension irons in place of J-hangers. In the 1920s, the V8 diagram became redundant when these FRUITS B were transferred to Y6 (cf also Y5). No further 21ft goods vans were built after the last of V7 in 1907, but there were later fish vans (S6/12). Unventilated 21ft vans were MINKS B, ventilated were MINKS C; vacuum fittings were irrelevant in the MINK codings.

About the same time that the first V8 was produced, a pattern 28ft 6in goods van appeared, with two doors per side, sequentially numbered 71945. Given V9, it had no vents, but was built with the vacuum brake (strictly 'converted' under L480). Underslung springing was used for the 20ft wheelbase. 100 more were built in 1906, the last 50 of which shortened the wheelbase to 18ft, giving

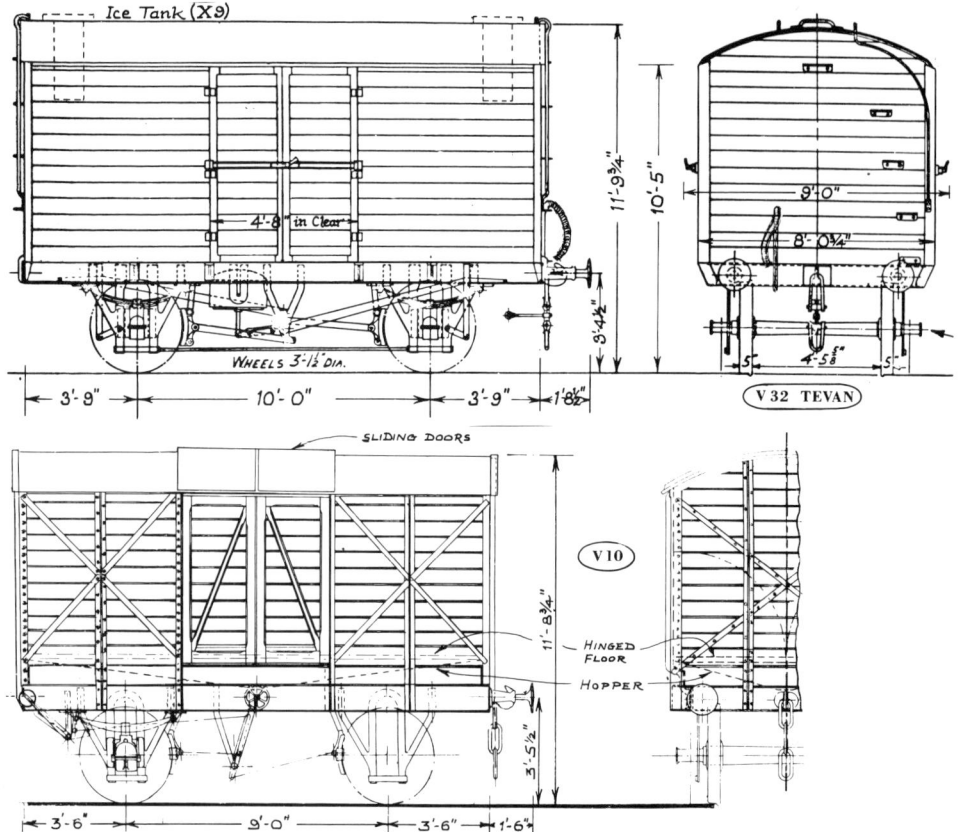

rise to the new diagram V11 (RWA Fig 46). All these MINKS D had twin-shuttered end louvres when built; a further 25 were built in 1911, with double V-hangers in place of the single V-hanger. No other 28ft 6in goods vans were thereafter built; thus no self-contained buffers occur on any of these long wheelbase vans. In common with the 16ft and other vans as built, no door catches were provided to prevent the doors swinging when open, nor were metal rubbing patches put on the doors. These features appeared in the 1920s.

Although of much bigger volume, the large vans were rated at 10 tons and remained so throughout their lives; it was not until the 1920s that the ordinary vans were uprated to 12 tons. V2 was a 'heavy' exception, but evidently van traffic was more bulky than heavy, as explained in Vol.1, p.19, and V2 was not continued. However, in 1931, 100 heavy fitted vans were built, 30ft over headstocks, 19ft 6in wheelbase to carry 20 tons, as part of GWR efforts to introduce higher capacity stock. These MINKS G were given V22 and may be considered to be an updated 28ft 6in design (Vol.1, p.18). The underslung underframe with short 3ft 6in springing resembled the contemporary DAMO A, except that the tierods were flat strip. It seems

that it was not clear originally whether to fit screw connections, but an entry in the lot book says 'Instanter decided by Mr Collett 22/10/30'. Large headed self-contained buffers were used as a safeguard against buffer locking, especially since the wagon could negotiate curves of 2 chains radius.

Other vehicles, whose origins lay elsewhere, were transferred to the V index in later years. V30 (see p.6) was the ex-cattle truck ALE wagons, fitted out in July 1939 for Messrs Guinness' traffic from Park Royal. Many of the very old 18ft over headstocks MEX trucks from W1 were converted. V31/2 were ex-MICA TEVAN vehicles for Messrs Lyons tea traffic from Greenford and Messrs J. S. Fry & Son's cocoa traffic from Keynsham. Also V35 (out of sequence) related to 650 SR non-vacuum vans with characteristic elliptic roof that were built at Ashford and absorbed into the GWR in 1942-3. BGW p.75 illustrates a plywood version; the GWR vans were the earlier combination of odd plank widths.

Also in the V group were bulk grain vans. L500 in 1905 built an experimental 20-ton convertible grain wagon, which was given V10. The body was like V5, but mounted on a 'double' chassis, with a

V32 ex-X9 TEVAN 'Van for special traffics'.
Photographed when converted in 1938 from 17ft 6in oh
dry ice MICA A built 1931. RCH fittings, Morton
brake. Other TEVANS written 'Return to J. S. FRY
& SON'S siding, Keynsham and Somerdale, GWR'.

V20 20-ton wooden grain hooper van convertible for
ordinary merchandise use. Photographed when built
1927. RCH short rib buffer, and axleboxes, but
rectangular works plate. Slotted link Morton brake,
toothed rack. Unusual door fittings. (RWA Figs 197/8).

V20 same vehicle but photographed in 1946. Doors
removed after reconversion from cement van. GRAIN is
World War II code.

sliding door in the roof. Versatility was aimed at
in the early days, so it had a false hinged floor on
top of the hopper, allowing the van to be used
also for merchandise traffic. More were intended
to be built under L531 but were cancelled, and
most grain continued to be carried in sacks in
ordinary MINKS or sheeted OPENS. The
Swindon wagon note book says that 60 sacks
of wheat were to be reckoned as 6 tons 12cwt, and
60 of barley as 5 tons 7cwt. No more was heard
of bulk grain traffic until the end of World War I,
when Order O.1154 of August 1918 was issued to
'make necessary alterations to 25 standard 16ft
covered goods vans for grain (drawing 56681)'.
No diagram was assigned however. Next was the
idea of fitting hoppers to 12-ton OPENS A (O25)
in 1925-7, and then in 1927 came the construction
of 12 new convertible 20-ton wooden grain vans
(V20). These were 21ft 6in over headstocks with a
rather short wheelbase, 10ft 6in. Why the 21ft or
22ft underframe and body of earlier vans was not
used is not clear except that the current Pole
coal wagons presented the basis of a versatile
20-ton underframe. Indeed, because of the hopper
extending below the underframe, the slotted link
Morton brake introduced on V20 pre-empted
that on N32. The doors on these grain vans were
much sturdier than normal, to take the side thrust
of the grain. Small windows were fitted high up on
the ends to see if the vehicle was loaded or empty.
The interior hopper ends could be swung down the
length of the wagon to allow their use as ordinary
vans, RWA Figs 197/8.

The vans had a chequered history. In the early
1930s, they were converted to bulk cement wagons
for the Aberthaw & Bristol Channel Cement Co,

V25 GRANO photographed when built 1935. RCH long
rib buffers and fittings. Disc wheels. Hopper wheel
chained up. V-hanger at control wheel not covered by
plate on earlier wagons. Inspection cover other side of
wagon. Some wagons marked to work between
Avonmouth and Uffculme, and for Coombe's traffic
(Showerton).

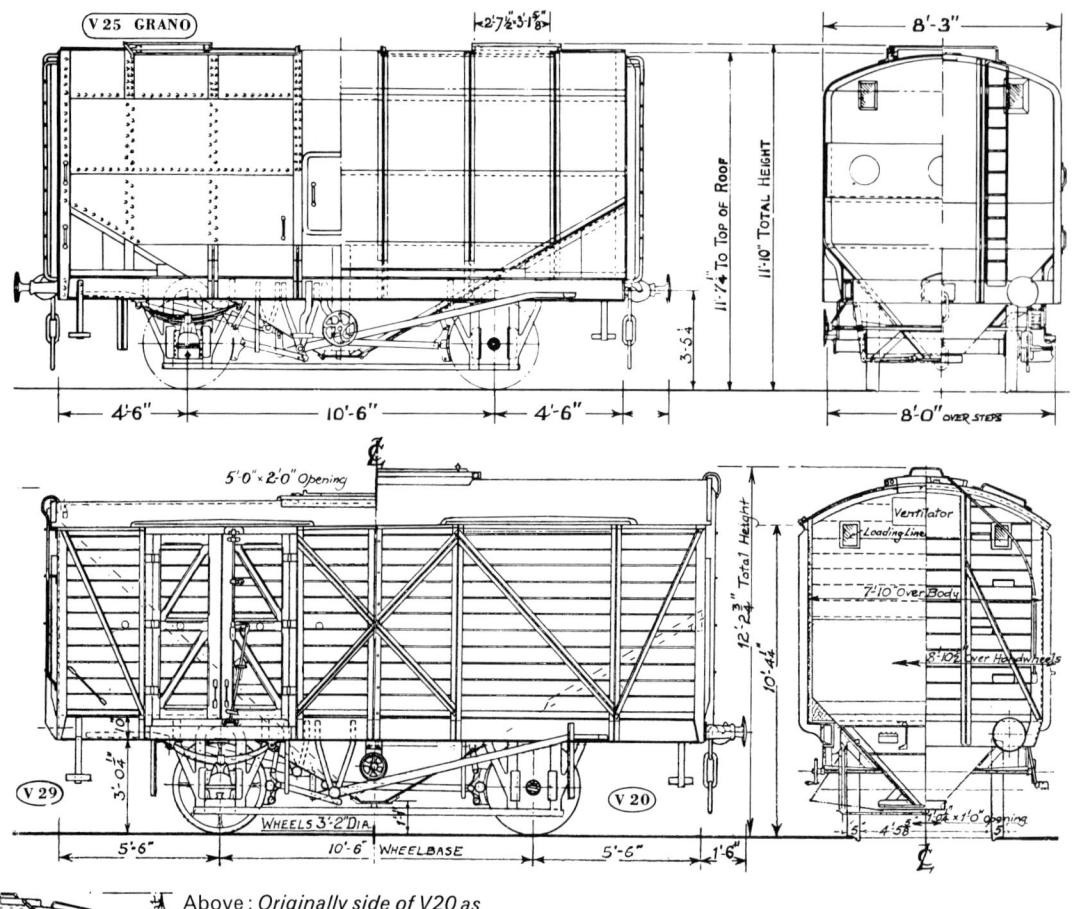

Above: *Originally side of V20 as shown for V29. Doors omitted later. End of V29 left.*

as the grain imports for which they were built had ceased. The interiors were lined with galvanised steel and had much steeper false hoppers, although externally only the provision of a flatter roof and the absence of the small glass windows at the loading line in the ends were evident (V29, out of sequence). Since the sloping floor cut high across the doorways, convertibility was abandoned and the doors screwed up. In the summer of 1939, 42243 was reconverted to a grain van and all others followed by the end of 1940. The high roof was restored, but the sides were completely rebuilt with

no doors, although the 1939 *Special Wagon Book* suggests otherwise.

In 1935-7 twelve steel grain hoppers coded GRANO were built on an essentially identical underframe to V20, except that they were 2ft shorter at 19ft 6in over headstocks (V25). Side inspection covers were provided on one side, but the vehicles were not convertible. Two openings with sliding covers were provided in the roof. The legend 'Door must be kept closed and completely fastened when loaded or empty', would greet anyone climbing the end ladders on to the roof; separate steps and commode rails were absent on this 'modern' design. The underfloor hopper opening had grown over the years from 6in by 4in on V10 and 1ft by 1ft 0¼in on V20/9 to 1ft 3in by 1ft 3in on V25. The handwheel controlling the opening was kept chained and locked. The open V-hanger for this wheel was replaced by a triangular plate on 42306 and the second lot.

The code name GRAIN was painted on V20 as a result of World War II reorganisation, but they were uncoded previously.

CHAPTER 21
W – CATTLE, AND PEDIGREE CATTLE, TRUCKS
(Mex, Beetle)

British Goods Wagons p.88 explains the historical quirk by which there were three standard sizes for cattle truck, as was reflected in the first three MEX designs in the W index. All were 8-ton wagons, but they differed in length: the large W1 (18ft inside, 11ft wheelbase) (Vol.1 p.73), the medium W2 (15ft 3in inside, *not* 15ft 6in, 9ft wheelbase) and the small W3 (13ft 6in inside, 8ft 6in wheelbase). Wagons of all three sizes had been built from the earliest days, indeed osL2 in 1869-70 related to 18ft cattle trucks, and Saltney built 50 such wagons on osL86. These were all outside-framed vehicles with wooden solebars. The first iron frames appeared on a medium cattle truck under osL198 in 1879 (osL199 fitted the first identical underframe to an outside-framed wooden goods van), and vehicles of this type with outside-framed bodies became W2 when the index was set up in 1905. None of these were produced after 1883, and the years 1887-8 saw the emergence of the channel-metal solebar cattle-wagon body design that was reproduced with variations into nationalisation. Although the lot list is not helpful, photographs show that osL433 built three distinct sample designs. Most wagons in the lot were small size (which became W3), but the second wagon 38102 had a large 18ft body, with early Armstrong vacuum brake equipment and Mansell wheels, and was one of the first 'goods' wagons to be given passenger equipment. It was an important trial vehicle for many reasons: the underframe fittings were modified for the MICAS and FRUITS which appeared in the following years (particularly concerning the length of the brake lever and the slotted guide on the brake cylinder),

and the body, with minor modifications but on a simple one-sided lever-brake unfitted chassis, was produced in large numbers in the 1890s to become W1, Vol.1, p.73. The first wagon in osL433, 38101, was also an important experimental vehicle, but one that was not pursued. It was an iron-bodied medium MEX built on the contemporary 16ft MINK/four-plank OPEN underframe; an experimental iron TOAD had been built the previous year.

As for 38102, the fitted concept was taken up on cattle trucks for pedigree animals, and the first BEETLES were essentially like 38102 with one extra plank to fill the gap between the sills and guard rail, and the rest of the sides louvred-in to the roof; like W1, these vehicles had a narrower door than the parent wagon, and additionally had the inspection/dung holes at floor level covered by sliding shutters. These 'special' cattle trucks built with the revised underframe were given W4 when

Bulb underframe.

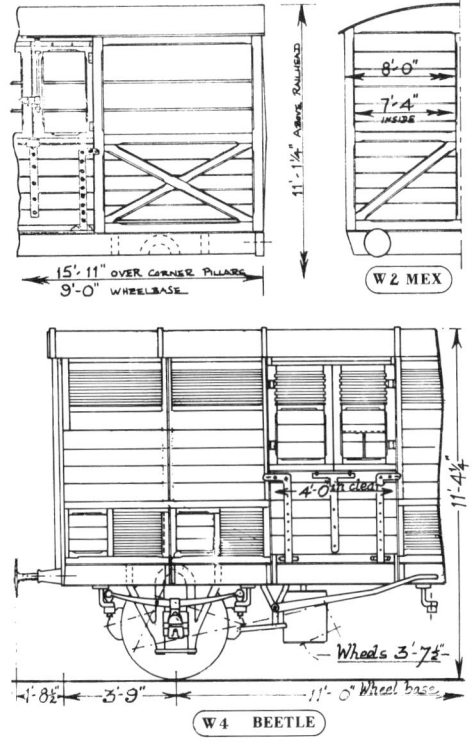

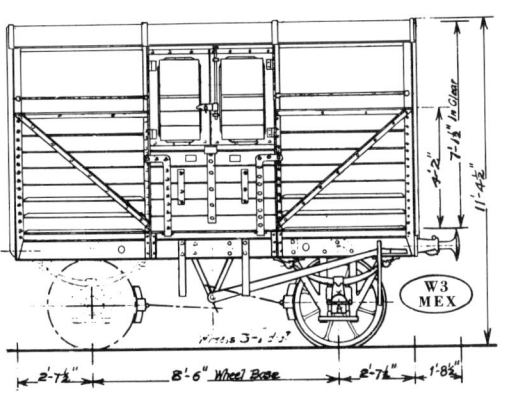

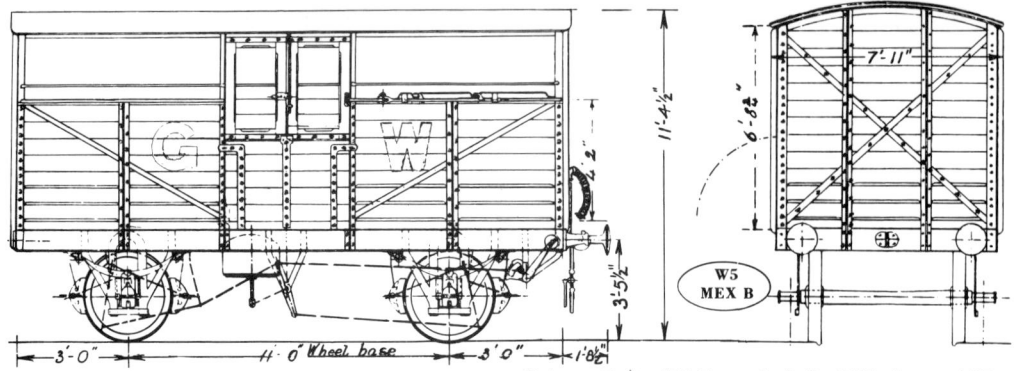

Above : *Later with wagon brake blocks. For W1 as built, substitute single side lever brake (some with offset V-hanger for standard length lever), two wagon shoes. Later DC I brake, screw couplings.*

Below : *End as V5 (but taller). For W6, shorten W7 to 23ft oh, 14ft wb, doors, and compartment as W7, rest reduced proportionately. For W13, as W7 with compartment mouldings sheeted over, oil lamps, ends as W8 (no slats). For W14, substitute central gas lamp on W13, ends as V22 (but no bonnets).*

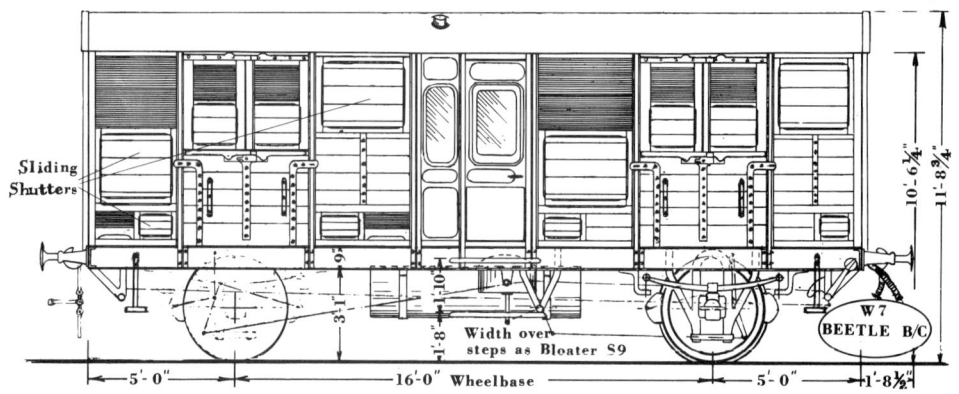

the index was set up; some were Westinghouse piped or dual braked, for which BEETLE A was used. The 1900 telegraph code book has the code BEETLE B for a vacuum-fitted goods special cattle truck, BEETLE and BEETLE A being called passenger vehicles. Presumably the original 'open' 38102-type were the B wagons, which were rebuilt with louvres and such like by 1907 when the BEETLE B code came to mean something different (see later).

In the development of ordinary goods cattle trucks, only the large variety was eventually built; but movable partitions were provided that sealed off the wagon into the size paid for by the consignor. The letters L, S and M painted below the sills indicated the size, and to prevent cheating the Wright-Marillier* locking device (patent 21, 101 of 1903) was fitted to GWR wagons, whereby the partitions could not be moved once the top swing doors were closed. W5 of 1902 was essentially W1 with (originally) DC II brakes, and subsequently after 1905 (moving piston) vacuum fittings. L494 was issued to convert many cattle wagons to the vacuum brake during the intro-

*F. G. Wright, Assistant for Works, Loco Carriage and Wagon Department; F. W. Marillier, Manager, Carriage and Wagon Works.

duction of fast fitted-freight trains and hence arose the MEX B code. Even before these rebuildings cattle truck buffers were often 1ft 8½in passenger type with screw couplings; likewise lamp irons were often provided. Since all new cattle truck construction was of MEX B vehicles, GWR cattle trucks were not in the common user pool immediately after the grouping. Flat cross-bracing on the ends of most vans gave way to modern end stays just before World War I, as shown in W8 of 1914, which also had higher pitch roofs instead of the low iron roofs of W1-5. W8 had a longer 11ft 6in wheelbase, with DC III vacuum brakes and self-contained buffing and drawgear. After the amalgamation, W8 was produced with the Morton brake (becoming W10) but with the detail body difference that the guard rail along the side of the wagon was 10in above the sills instead of the previous 6in (in the doorway it remained at 6in).

The first cattle trucks with RCH fittings were W11: the body was essentially W10, but the design used an RCH 12-ton underframe 'modified to suit the GW body'. Of interest was the use of 1ft 6in RCH buffers that were packed out 2⅛in (and the trimmers adjusted) 'necessary to suit the coupling'. The variation in the final MEX B diagram W12 was that the top side doors were

W4 BEETLE B (original code). Prototype passenger cattle wagon, photographed when built 1888. Armstrong vacuum, brake rods outside central V-hanger. Extra long lever thus jogged at hanger. Central white stripe to indicate brake lever side of wagon. Compensating rods in front of 8 by 3½in OK boxes. 1ft 8½in buffers, Mansell wheels. Safety chains. Hinged bar screw coupling. Body W1 proportions, subsequently louvred-in for W4 diagram with narrower door, and offset V-hangers, shorter lever, as on X1, Y2 etc (cf K2).

W8 MEX B built and photographed 1913. High roof, stayed ends, DC III vacuum brake, GWR s/c buffers. Wide door stops.

vertically planked-in level with the cattle rail, a pin hole allowing opening, but otherwise it was W11. About this time in 1929, a proposal was put forward for double deck sheep wagons with end loading; from the Swindon proposed drawing, the vehicles evidently were modified W11 bodies, but nothing came of the idea.

With the loss of cattle traffic to the roads some of W1 were reclassified in 1939 for Guinness traffic (ALE wagons, V30). Amongst those transferred were some from osL464/76/651 which had been built 6in shorter at 18ft over headstocks. At least 228 of these 18ft wagons remained in service in 1935. Similarly, 130 wagons from W10 were converted to goods fruit vans in 1939, becoming Y10 (see Y group). The side stanchions were left alone, so that the door, patterned after the contemporary vertical plank door of goods vans, was rather narrow at 4ft in clear; sometimes the drop door stops were left on the underframe.

Attendants travelling with pedigree cattle would occupy seats in the train, or on goods trains might

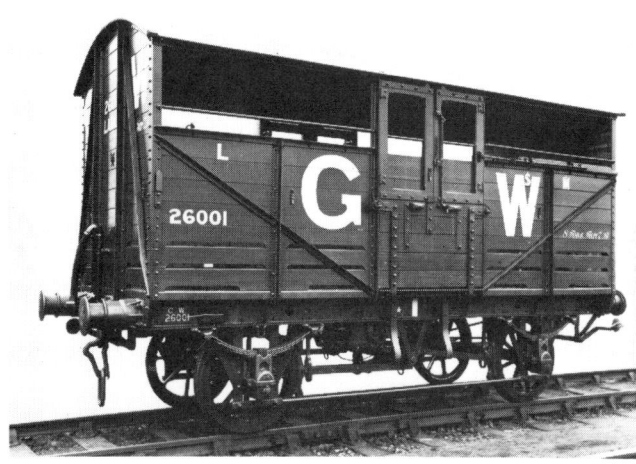

Prototype iron cattle truck photographed when built 1888. 16ft u/f, height and proportions as V6 iron mink (including 4ft 10½in in clear door). Grease boxes, one-sided lever brake, vertical rod inner V-hangers. Note spring bearing shoes. No more constructed.

W2 MEX medium cattle truck. Built 1883, photographed in July 1904 during 25½in large GW painting trials, to illustrate how lettering would be accommodated to those wagons without sufficient flat area to take full-size letters. Note serif on crossbar of G. End livery, a minor variation from earlier 5in block capital style using 'GW' under eaves on end instead of 'GWR' and omitting the 'To carry' with load, and quarters in tare figure on the end panels, which on the left read MEDIUM with 26481 below, and on the right 8 Tons with Tare 6—4 below. This end style not adopted, and large GW in end panels used even after 1922 (then with 16in capitals). Chalk marks for painter still on wagon. Wright-Marillier partition device of 1903 fitted. Bulb underframe, horse hook, grease boxes, one-sided lever brake, rigid connection to plain shoes, non-vacuum yet screw coupling. Round base buffers. Body construction and framing similar to wooden goods vans of last quarter of 19th century.

(W. O. Steel Collection)

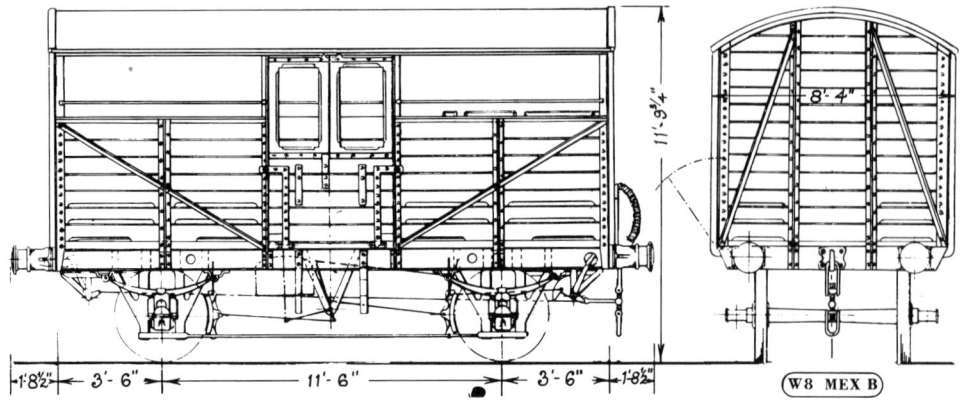

(W8 MEX B)

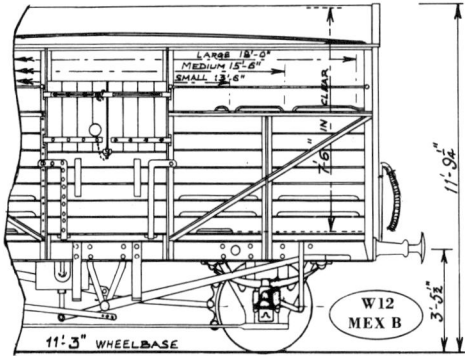

W12
MEX B

11'·3" WHEELBASE

Above : *For W10, body as W8 (no slats in end, repositioned rails), substitute Morton brake. For W11 as W10, shorten wb to 11ft 3in, RCH fittings and packed out buffers.*

travel with the guard. To allow closer watch to be kept on the animals, the dual fitted BEETLE B (W6) of 1907 contained a central grooms' compartment, with boxes on either side capable of stabling two beasts each. This 'one-off' sample wagon was much improved over W4, fodder racks and water troughs (with a tank in the roof) being provided. Slightly bigger accommodation was allowed for the animals in the W7 design, subsequently produced in quantity. The 9ft 4¾in lengthwise size of each cattle box was increased to 10ft 10¾in, thus making W7 26ft over headstocks as opposed to the 23ft of W6. Full carriage fittings were provided on these BEETLES B, such as incandescent gas lighting with a 7ft by 1ft 6in diameter gas reservoir under the wagon, steam heating and communication chain connections. L639 of 1909-10 built 20 vehicles, and then ten further wagons were built in 1926 under carriage lot 1380 (after 1915, brown vehicles were not built under goods lots). Because of the obsolescence of air brakes after the grouping, this latter lot had only vacuum brakes, and as such were the first BEETLES C. They also differed from the earlier vehicles in that mouldings around the attendants' compartment and door gave way to plain sheeted sides, and the ends had modern stays

like W8 instead of crossbracing. A detail change from the 1910 vehicles was that the roof water tank was omitted. Similar pasenger cattle boxes were built to diagrams W13 (1931) and W14 (1937). On W13 the doors omitted top ventilators and on W14 the end stays went only three quarters of the way up akin to goods vans (but without bonnets of course). Two oil lamps were provided for lighting on W13, but W14 reverted to gas Certain underframe differences occur on W14; for example the 1ft 8½in buffers are the self-contained variety with large heads, rather than the carriage type used on all others, and the side stanchions are not tucked under the solebars. The six-wheel vehicle preserved by the GWS at Didcot is a Western Region, not GWR design.

All cattle wagons were built and operated according to government regulations (Appendix 6 of BGW) with battens on the floor and drop doors. Floors were often tarred or creosoted because of the corrosive nature of cattle dung. Although the 1904 law forbade open cattle trucks, the exigencies of World War I forced a waiver of the regulations, for 410 wagons of O2/10 were modified for military horse traffic by the addition of two wooden rails above the open sides, like coke rails. 28023 for example was thus converted for the War Department in 1914; all were reconverted to 4ft 3in merchandise wagons early in 1919.

Diagram W9 was issued in 1923 for two wagons for the Vale of Rheidol narrow-gauge branch, which had been taken over in the grouping. Later, under Order 3529 of 1937, these vehicles were converted from 1ft 11½in to 2ft 6in gauge for use on the Welshpool & Llanfair branch.

CHAPTER 22
X – MEAT VANS
(Micas)

Early in this century beef was imported into this country 'on the hoof' from America, whereas because of longer journeys, Australian and New Zealand mutton was imported frozen. The first refrigerator ships were introduced in the early 1880s. Much of the live beef came into Birkenhead, and after the cattle had recovered their condition in lairage for some days, they were slaughtered and the butched carcasses 'chilled'. Only ventilation was required to keep such chilled meat cool on its way to market, the time between abbatoir and Smithfield, say, being less than 22 hours, of which 12 were in cooling and about 7 in the train journey to London; frozen meat on the other hand clearly required refrigerator meat vans. These different needs gave rise on the GWR to the different types of MICA meat vans. For loading purposes, the reference weights of carcasses were 180lb for the fore-or hind-quarters of beef and 56lb for mutton. Typically 60 large hooks would be provided in the roofs of MICAS for hanging the meat.

There were wooden underframe standard gauge meat vans with outside-framed wooden bodies dating from the late 1870s, such as osL169/230, numbers 28866-940. They were similar to the outside framed goods vans of the period. A broad gauge example of a ventilated meat van is seen on the back dust-jacket of RW. An early experiment in refrigerator vans is shown here by the photograph of van 28657. Minute No. 22 of the GWR Board Meeting of 2.7.1874 states that 'an arrangement has been made with Capt. George E. Acklom by which he is to be allowed to fit up one of the company's vans with his patent refrigerating apparatus and in consideration of which the company are to have licence to fit any of their vans or trucks for the conveyance of Meat, Fish and Perishable Goods with such apparatus without he payment of any royalty in respect of the same'. In the following year £252. 5. 3d was authorised for fitting up a covered van (minute No. 8 of GWR Board Meeting on 17.2.1875). The van chosen was the ex-broad gauge vehicle 2975, which had 'narrow gauge' number 28657. It was reported in the *Railway Flypaper* magazine for November 1874 (p.347) that a van equipped with Capt. Acklom's Refrigerator Patent Apparatus was loaded with a quantity of slaughtered ox, despatched to Aberdeen at the request of local butchers and interested parties in London and Aberdeen, on 19 October 1874, returned to London on 28 October and meat examined and found to be in excellent condition, fit for sale and none the worse for its journey. The GW van was condemned in June 1893, the photograph evidently being taken at scrapping.

The GW also possessed refrigerator vans (unventilated) for chilled meat having inherited 13 wooden framed goods meat vans from the

For X4, as X2 with roof doors (as X7), end steps and rails (slightly different layout from X7), Armstrong lever replaced by DC quadrant (V-hanger same place).

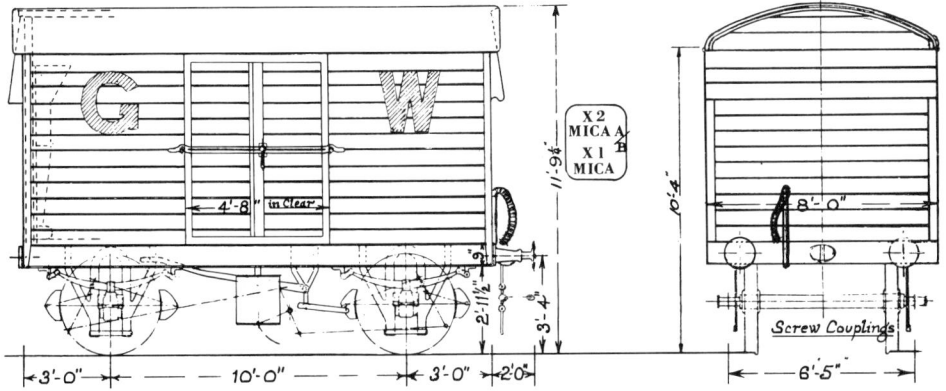

South Wales Railway (built at Swansea Wagon Works). A drawing of these diagonally planked non-GW wagons of 1880s vintage which subsequently became X3 in the GW Wagon Diagram Index is given in RW p.134, and photograph in RWA Fig 202. Although the SWR was absorbed into the GWR in 1863, its goods vehicles were not given GW numbers until the 1890s, which accounts for the particular range of then current serials used (i.e. 47738-50; former Pembroke & Tenby outside framed wooden solebar covered goods vans also received 47xxx numbers.) The means of refrigeration in these vans was used on subsequent GW designed refrigerator meat vans. All the SWR vans had been condemned by 1910.

The earliest GW vans for meat traffic that survived into modern times date from the late 1880s. These were so-called 'ordinary' meat vans (with the MICA code, no suffix) which were merely ventilated for carrying fresh meat and had no provision for ice. The characteristics of this design (subsequently diagrammed as X1) that followed on most other GW meat vans were horizontal double-cased tongued and grooved planking with flush-fitting doors and no exterior bracing on sides or ends (cf ASMO, DAMO vertically planked in later years). Ventilation was provided by hinged wooden bonnet vents across the top of the ends of the vans, the sides of the vents being wood on X1 with a fancy scallop cut out. Also in the sides of these first MICAS were 1ft 8in long ventilation slots, backed by perforated zinc plates, in the third plank down from the roof; in later years these were filled in. The first fifty vans of X1 were built as half of the 1889 osL490 of 100 vehicles, the other fifty being Y2 FRUITS.

The first GW-designed refrigerator vans for chilled meat were introduced in 1897: they had ice bunkers inside the van at each end, which in the original type were filled from *inside* the wagons (cf SWR vans X3). One lot of these vans (L177) had neither end ventilators nor side louvres, and like the SWR vans, were coded MICA A, i.e. refrigerated but non-ventilated. Strictly there was no X diagram which subsequently described these GW vehicles (sketches of which are given in the SWB's of both 1932 (p.29) and 1939 (second drawing, p.34)), although L177 is normally attributed to X2. X2 in fact were all the other lots which had ventilators, and were more or less like X1 in external appearance except that the side profile of the end ventilators was straight-bottomed without the scallop of X1. In these vans there was a 3in airspace between the double body sheeting, giving some insulation. Interior ice bunkers were built in nearly all of X2, so with their ventilators, these vans had the capability of being either ordinary ventilated vans (the ice bunkers not being used) or refrigerator vans (using the ice with the vents shut off). They received the code MICA B, indicating their convertability. An entry in the lot book against L192 states 'Mr Thomas says to be marked like refrigerators so ice can be put in if neccessary.' The bodies of MICAS A/B were painted white with red lettering; ordinary MICAS were wagon grey.

The X2 MICAS B were Lots 139/52/92/333 and part of L429, the first ten vehicles of which (59791-800) omitted the interior ice tanks and were thus simply MICAS (RWA, Fig 205). These latter were nevertheless included in X2 when the diagram index was set up. Error occurred in the compilation of the X2 index in Vol.1 of this book: details given here are correct, i.e. only L429 (part) being MICAS, and the rest MICAS B. The non-diagrammed MICAS A (with no end vents) were only the ten vehicles of L177, serial numbers

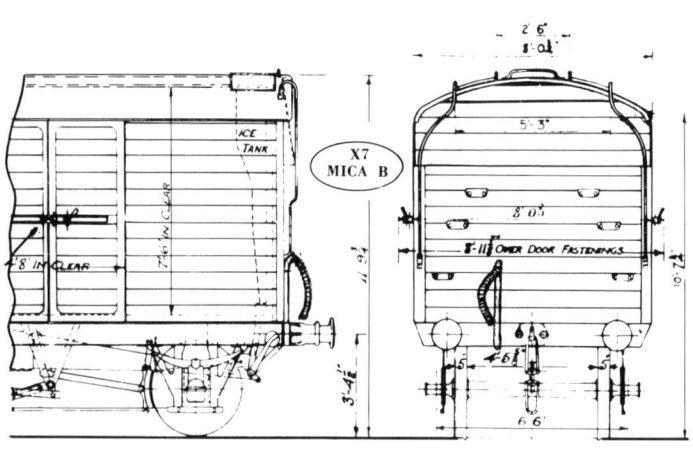

For X5, substitute GWR laminated sprung buffers, swinging bar door fasteners. For X8, substitute Morton brake. For X10, substitute dry-ice tanks as on V32 (X9).

Trial meat van fitted with Acklom's refrigeration apparatus in 1875. Ex-broad gauge van number 2975; photographed at scrapping in 1894. Wooden underframe, springs inside axleguards, grease boxes, horse hook. Two wooden brake blocks, arched lever one side, pinned rack. Five-link coupling, buffers packed out. Open spoke wheels. Outside framework of body (with horizontal planking) typical of wooden goods vans before iron MINKS. Grills for ventilation: air passed through layers of felt moistened by water (handpump connected to rectangular tank beneath wagon). Other meat vans (ventilated, not refrigerated), numbered 28xxx built 1870–80; all scrapped by about 1910, and not given place in X index.

X1 MICA built 1899 photographed 1941 with small lettering and changed telegraph code. Long lever brake (cf Y2) altered to cross-cornered DC quadrant at ends, but Armstrong sliding cylinder vacuum fittings retained, clasp shoes. Mansell wheels. 8 by 4¾in OK boxes 2ft carriage buffers. L-spring hangers. No special sealing device on doors. Scalloped vent, not adjustable. Miniature T-stanchions on buffer beam.

47811-18/908/65. Also in Chapter 2 of Volume 1, the MICA B code starts of course 10 years earlier than stated, viz: 1897 not 1907, and the MICA A code applies to diagram X2 as well. It is clear that when the diagram index was drawn up in 1905, the newer vehicles of X2 should have preceded X1.

From about 1907, the ice tanks in new constructions were extended upwards and rubber-sealed trap doors were provided in the roof for filling from outside; steps and commode rails appear for the first time on the ends of the wagons. The tanks were divided into compartments, the top for ice and salt, the bottom for retaining brine, with cocks for draining. The body of this new design (X4) was X2, with the addition of steps, rails and roof hatches. As with other meat vans, the insides except the roof were lined with zinc, and the floors either zinc or asphalt.

The 16ft underframe of X1/2/4 were the 'classic' early fitted design: coach wheels, Armstrong vacuum brake, clasp shoes, 2ft buffers, screw

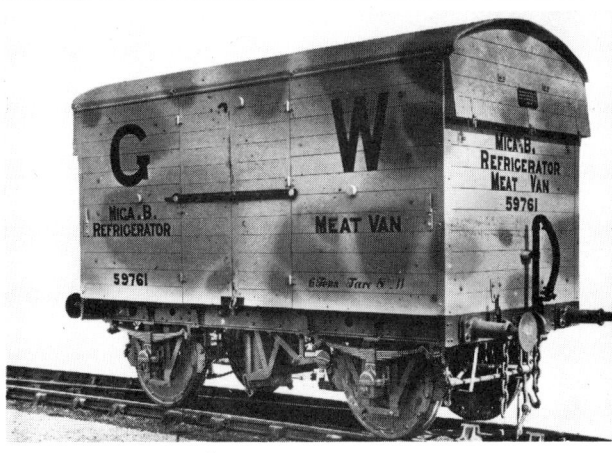

X2 MICA B built 1904, photographed 1908. Interior ice tanks, filled from inside the wagon (no steps, rails or roof hatches). X4 is identical with addition of roof hatches) etc. Early X2 had long lever brake instead of DC lever. Armstrong vacuum, Mansell wheels, clasp shoes. Swing bar for sealing doors. Lamp irons on sides. Coach buffers and safety chains. Early screw coupling with hinged bar. Plate on ventilator says 'Shutter to be CLOSED when used as REFRIGERATOR van. To be left OPEN when used as MEAT van'. White livery, red lettering. G has serifs from earliest 25in alphabet. Miniature T-stanchions (cf RWA Figs 203/5).

X7 MICA B built 1921 photographed 1940 with small lettering and new code. Self-contained buffing and drawgear. DC III brake, wagon wheels and shoes, GW axleboxes. Screw coupling with solid bar. New pattern door fasteners, door stops on curb plank. Roof hatches and commode handles. Plate on ventilator like X2. Miniature T-stanchions on ends. Numbers 59801-20 in same lot had been used previously by Y2 before latter given van numbers as brown vehicles.

connections and long 10ft wheelbase. X1/2 had identical chassis; X4 had the long lever replaced by a DC quadrant in the Armstrong V-hanger location, and had the newer fixed-cylinder arrangement. The vacuum fittings on X1/2 were changed eventually, some looking like X4, others having a DC III system.

In 1912, freight wheels and axleguards replaced the Mansell coach wheels, as shown on X5, the body remaining as before with a slight realignment of end rails; the height of the vans remained at about 11ft 9in, and the rather low buffer height of 3ft 4½in was perpetuated, as on all other GWR MICAS. Also on X5, the vacuum layout was completely DC Mark III. Ten vans were built under L684 to this design, and many of the last lot (630) of X4 were also built this way (such as 79679). The last use of 2ft coach buffers occurred on this design (cf V8, Y1/2/6).

No new MICAS were constructed during World War I, but a need for additional insulated vans was met by converting 300 10-ton covered goods wagons from V16 on Order 1060 of May 1918. Typical meat hooks were provided in the roof, but since these were not vacuum fitted, they were 'goods' MICAS A. Their life as meat vans (X6) was comparatively short, and they were reconverted into covered goods vans from June 1921 on Order 1510; some however were directly rebuilt to banana vans (Y4).

X7 of 1921 was X5 with 1ft 8½in GWR 'C-type' self-contained buffers. The method of door-locking was new, bars fixed to the doors replacing the 'swinging bars' of earlier designs, and rubber seals were provided. Three of these vans were uprated to 10 tons in April 1922 for conveying ice from Slough to Paddington (Order F420). X8 of 1923-6 was X7 with the Morton brake, but otherwise with GWR fittings.

X10 is the next logical diagram. Here, the wagon was outwardly X8 with no end vents, but a fundamental change had occurred with the insulation and the refrigerant. Dry-ice was available commercially under such brand names as 'Drikold', and the two experimental vehicles built in 1927-9 to the X10 diagram had small 'feeder' ice tanks hanging from the roof, whereby the refrigerant could be introduced into the vehicle. 'Onazote' or 'Alfol' cork insulating material was stuffed into the airspaces at sides, floor and roof. These materials were also being used in contemporary carriage construction. Fifty-four similar MICAS A were built in 1930 to the X9 design. Again, no vents were provided (to prevent excessive sublimation of the solid carbon dioxide), and for the only time, a 17ft 6in RCH underframe was employed on GWR meat vans. The body was thus a stretched-out version of X10, keeping the same 4ft 8in in clear doors, the only modification being the provision of merely one rearranged commode rail and set of steps at each end.

Because so much meat traffic later went to the roads, or at least was rail carried in GWR containers such as types M (ventilated) E and FX (insulated, Drikold refrigerated), many meat vans were converted in 1938 to TEVANS (V31/2) for the conveyance of tea and chocolate. Nevertheless representatives from all the X-group (except X3/6) passed into nationalisation. Many of the old vans were provided with interior slings for dry-ice bags in 1942.

The MICA codes disappeared in World War II. The ventilated vans were then coded MEAT, and the words 'MEAT (for fresh meat)' were painted on the van side. Both the MICAS A and B were given the code word INS, but were distinguished by the words INSUL-MEAT and VENT-INSUL-MEAT respectively painted on their sides.

CHAPTER 23
Y – FRUIT VANS
(Fruits)

1890 was the first ventilated vans built specifically for the conveyance of fruit. Fitted to run in passenger trains, with 10ft wheelbase, Mansell wheels and Armstrong vacuum (and some with Westinghouse pipes), over 100 vehicles to the Y2 design had been built by 1900. In overall body size, they were the progenitors of the 'low' ten-plank wooden goods vans V5, built 1902-4. The ventilating arrangement on Y2 was odd, and

not perpetuated later; the floor planks had gaps between them open to the track! To give exit for the air current, small single louvres were bevelled in between the rows of planking, together with central end bonnet vents. One hundred taller wagons, akin to the taller of V4, were built in 1904-5, and became Y1. The underframe, like Y2, was similar to the MICAS of that time, but the body had louvres between every plank, in the ends

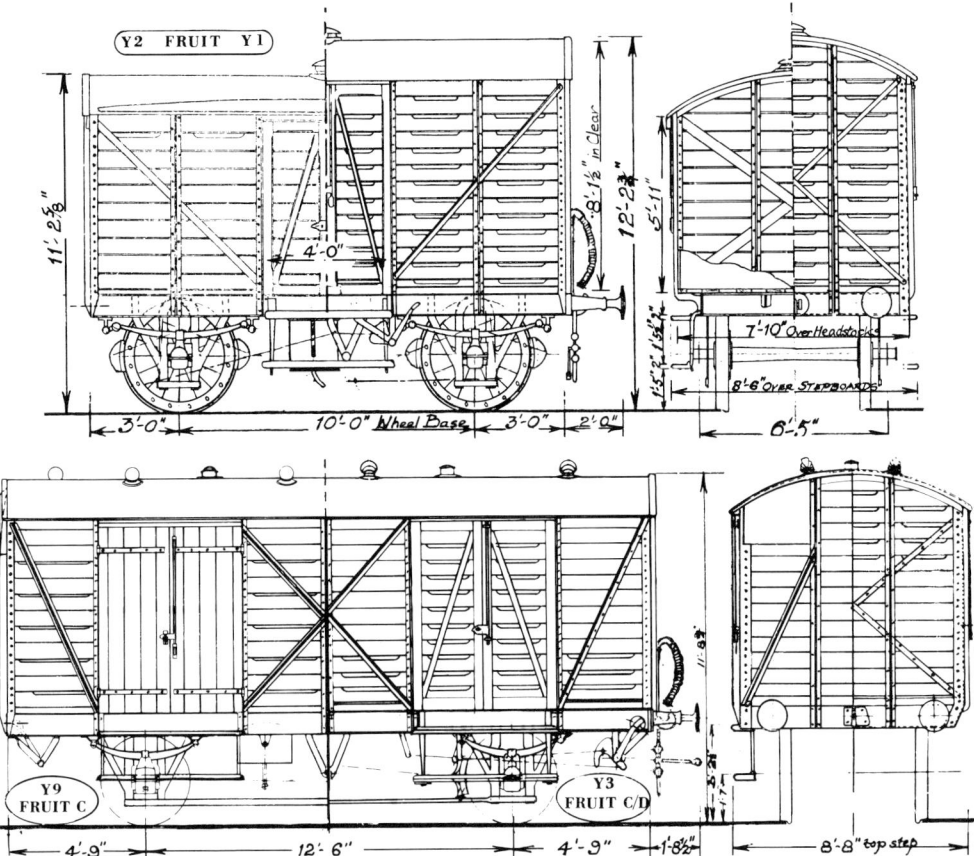

as well as the sides; no end bonnet vents were present, nor was the floor open. Rows of galvanised wire trays on wood frames were provided for stacking the punnets of fruit. Both Y1/2 had 6-ton capacity; a larger 10-ton design was produced in 1911-12, which became Y3. These vehicles were 22ft over headstocks, 12ft 6in wheelbase, and were essentially the same design as the 21ft, 12ft wheelbase V3/7 (including the use of 3ft 1in wheels and wagon springs), but by this time the vans had DC III brakes and normal wagon brake shoes. An indicator of passenger train working was the use of 2ft carriage buffers, however, and also of the 100 vans ordered through L667-8, 75 were built under the first lot with dual brake systems (vacuum and Westinghouse). The ventilating arrangements on Y3 were single louvres between the planks, but not everywhere as in Y1; the doors had slats all over, but then only the bottom three planks on the sides, and the top four planks on the ends, had louvres. Gauze inside all these vans behind the louvres protected the contents from dust and engine smuts. Like Y1/2, Y3 were lighted, but with incandescent gas burners rather than oil lamps as the former vehicles.

The traffic with which these and later vans were principally associated was the conveyance of apples, plums and other fruit from the Worcester area, and at least 200 vans were specially earmarked for return to Pershore, Aldington Siding, Littleton and Badsey, and Evesham. They also had ample use for the West Country broccoli and Channel Islands tomato traffic.

The importation of bananas, and ideas of ripening them by steam heating en route to the marketplace, gained ground in the first decade of this century. In 1884, 10,000 bunches represented the annual import of bananas into Britain; in 1910, Messrs Elders and Fyffes Ltd imported more than 200,000 bunches in a single week. At Avonmouth Docks, three lines of vans were loaded simultaneously, gangways being left between the wagons for access. As many as 400 or 500 vans used to be employed for a single cargo. To prevent ill effects to the bananas, they were stowed in the vans on beds of straw and had to be kept at an even temperature of some 56°F. Railway vans therefore eventually came to be fitted with ventilators and steam heating apparatus, for use in summer and winter respectively.

Y2 photographed as built 1890 in grey livery. Coach buffers and wheels, Armstrong brake, long lever, clasp shoes, safety chains. Central white stripe. Door design became pattern for later wooden MINKS; no catches to hold door open. Oil lamp ventilator in roof (RWA Fig 52 for brown livery).

Y3 FRUIT D (original meaning of code) built 1911 as 85960, photographed 1920 in brown livery van number 2460. 22ft version of 21ft V3/7, flat cross bracing on ends. Shell roof vents and gas lighting, tank pressure gauge and lever near V-hanger. 1ft 8½in buffers. DC III brake, GW 8 by 4in boxes. Dual fitted, Westinghouse and vacuum (W star bottom right door). 'Return to Gloucester immediately'.

The first van design specially built for banana traffic was the 10-ton, 21ft over headstocks, 12ft wheelbase V8 number 79144 of 1905. As explained in Chapter 20, it was the first ever vehicle to have adjustable shuttered louvres in an attempt to control the ripening. Twenty similar vans followed under L517, and L576 of 1907-8 built 180 more 10-ton banana vans, numbered 82521-700. These latter wagons were basically the contemporary goods vans, V12, with the addition of a steam heating pipe, and remained in V12 at that time, although the 15 V12-type fish vans were given their own diagram V13.

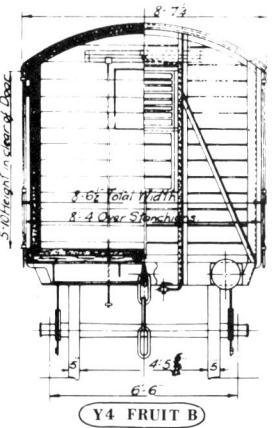

Y4 FRUIT B

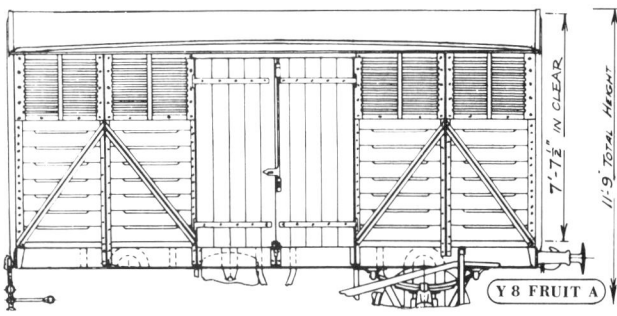

Y 8 FRUIT A

U/f and end as V22.

For Y7, put louvre shutter in left hand panel, flat cross bracing between end stanchions.

All these banana vans, although coded FRUITS B, were not originally assigned places in the Y index and they remained with the goods vans. That is why the 22ft vans took Y3, even though the above older FRUITS B were already running. The FRUIT code itself referred to the 16ft over head-stocks passenger vans (Y1/2), and FRUIT A to the Westinghouse piped (or completely dual fitted) vans in Y2. The codes FRUITS C/D related to the Y3 vehicles, C being the vacuum vans, D dual fitted vans. Confusion has arisen over the FRUIT D code because after it became obsolete when the airbrake fittings were removed circa 1930 (and the former D vans became C), it was re-used in 1939 for a longer fruit van design Y11. Likewise the FRUIT A code was re-used for 'goods fruit vans' after 1930 (downgraded Y1 and Y8 later).

At the end of World War I 300 V16 vans were converted into insulated vans (X6) and all were intended to be put back into merchandise stock in 1921; soon afterwards, however, Order F.428 of October 1922 took those that had not yet been reconverted and made them into banana vans. They were still insulated, and they received steam pipes and a single shuttered louvre ventilator between the stanchions at each end (the diagonal angle bracing remained on the ends from the time when, as V16, they had bonnet vents; these were blanked off when X6, so a central adjustable louvre was the simplest method of ventilation). A flat drop pin was provided at the bottom of each door, together with eyebolts for sealing up the vehicle. Vacuum fittings were added and also instanter couplings. The Swindon drawing shows three-link couplings which seems in error; it is likely that the X6 drawing was merely updated. These converted

banana vans were the first to be entered into the Y index as Y4. In the early part of 1925, 47 similar vehicles were constructed new under L403. The 16ft bodies were the same (the lots refer to print number 66508 which was associated with Y4), but the underframes had Morton brakes and apart from the central end louvres these wagons looked like the contemporary V18. However, they were not given a new index number at that time, and they were lumped in with Y4. They eventually appeared as Y12 in 1940, the discovery that they did not have DC III brakes presumably occurring during the post-1939 DC brake 'sort-out'. The Y5 diagram was taken by the 180 FRUITS B built in 1907-8 which had remained in V12. The reason for the transfer to the Y index was Order F441 which insulated them. These first steam-heated FRUITS B, together with V8, were not insulated, and presumably the success of the ex-X6 FRUITS B in conserving heat suggested the change. Soon, order F448 of November 1925 also insulated all of V8, which became Y6. The effects of the insulation were seen by a test run at Swindon in June 1934 when with steam at about 70psi and the outside temperature at 72°F, the inside of a banana van kept at 116°F for many hours.

Yet more banana vans were built in 1929-30, 100 taking Y7, and were essentially the V21 17ft 6in over headstocks RCH underframe design with vertically planked doors. Shuttered end louvres were provided as in Y4/12, but this time in the left-hand end panel, close to the wagon side; the arrangement allowed easier access from loading docks. Flat cross-bracing was used only between the end stanchions. This was the final banana van design built by the GWR.

Curiously, the first goods fruit vans (FRUITS A) built as such, rather than being downgraded

Y9 FRUIT C built 1937, photographed 1939 with GWR totem. Brown livery. Modernised Y3 design, conventional bonnet vents. 9 by 4¼in RCH axleboxes and long rib buffers. Oval works plate, 'standard' omitted. Disc wheels. The last new design with DC III brakes. Steam connection.

Y12 FRUIT B built 1925, photographed about same time. 16ft body of Y4, but Morton brake, RCH 9 by 4¼in axleboxes. GW self-contained buffers, slotted trunnion for vacuum cylinder. Central end shuttered louvre ventilator. Steam cock handle under works plate, instructions above, below running number. White disc, bunch of bananas. Hook catch lock under door.

passenger fruit vans (such as Y1 in later years), were the 200 12-ton Y8 wagons of L1270 in 1937-8, introduced for the Guernsey tomato traffic which then involved over 2½ million baskets (some 16,500 tons) per year. Based upon V23, from which end view they are undistinguishable, their characteristic looks are in the side elevation (RWA Fig 53). There is a decided similarity with cattle trucks, such as W4. A year later indeed L1363 converted 130 MEX B of the W10 diagram into 8-ton FRUIT A, by boarding up the rails with slatted planks, providing extra louvres between the existing lower boards and substituting a vertically planked door (4ft wide in clear, and therefore oddly narrow in comparison with the standard 5ft or so in clear). The end ventilation on these ex W10 FRUITS A, which were given the diagram Y10, had to be one central bonnet vent, because of cattle truck end diagonals going right up to the

roof. The ends were also covered with proofed canvas. It seems likely that the conceptual background to Y8 came from an idea that a ventilated van might readily be adapted from the cattle truck style. Y8 were, however, of definitely new construction, with a standard width door, and certainly Y10 were not put through the shops until later.

An updated 22ft over headstocks design was produced in 1937-9, which was really Y3 with fewer side louvres, vertically planked doors and end bonnet vents. The 50 vehicles, built under carriage lots 1601/34, took the FRUIT C code, and were the last GWR wagons to be designed with a DC brake (Y9). As passenger fruit vans, they had gas lamps and removable galvanised wire trays. Y9 were the only examples of 'modernised' versions of the pre-World War I 21ft/22ft over headstocks vans. Likewise, a modernised 28ft 6in design was the largest, and also the last, fruit van built by the GWR (Y11). Carriage lot 1649 built 50 such wagons between 1939 and 1941, which were like Y9 but with three doors per side. In that, they differed from say V11, but were much like BLOATERS S8-11 in size, except that the doors were hinged rather than sliding, and were of course by this time vertically planked. The Morton brake arrangements were new however. The longest wheelbase wagons using Morton brakes up to that time had been the 12ft wheelbase locomotive coal and mineral wagons (N23 et seq), some later versions of which had a slotted link clutch arrangement. A single V-hanger was impracticable on the 18ft wheelbase of Y11, so the cam clutch system was reverted to, with a connecting rod between two V-hangers, from which two normal size Morton hand levers departed. (A similar arrangement was

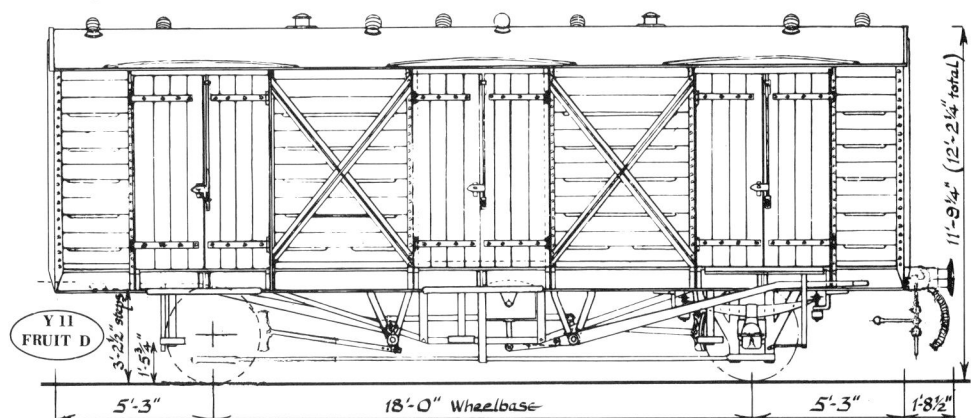

Y 11
FRUIT D

later used on the 19ft 6in wheelbase TUBES O41.)
Although of larger capacity than other fruit vans,
these FRUITS D (second use of code) were still
rated at 10 tons.

During World War II, the codes changed as
follows: the passenger fruit vans coded FRUIT,
FRUIT C/D became PASFRUIT, PASFRUIT
C/D, and the goods fruit vans, formerly FRUIT
A, lost the suffix to become merely FRUIT.
Banana vans were called by that name, the FRUIT
B code disappearing.

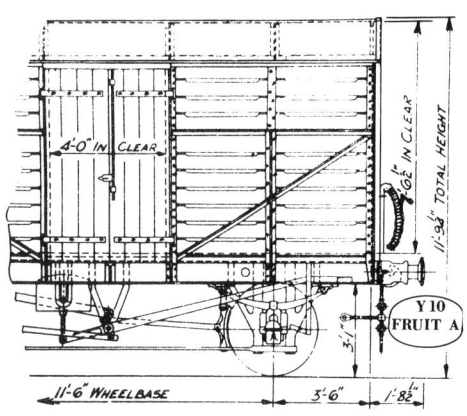

Y 10
FRUIT A

CHAPTER 24
Z – GUNPOWDER VANS
(Cones)

The construction of gunpowder vans was speci-
fied most closely. The body and roof were built of
steel plates; the inside was cased with boards,
secured with brass screws. The floor, sides, ends
and doors up to a height of several feet were lined
with lead sheets, fastened to the wood casing with
flat-headed copper nails. In opening or closing the
doors, there was no contact of steel with steel to
cause sparks, as hinges, locks, bolts and other
fittings were of brass. All the metal was well
coated with white-lead oil paint, where it came in
contact with wood. Vans had no ventilators and
were absolutely dust-proof. Inside the vans special
boots to prevent sparks were hung for the men and
special regulations about operation were posted. In
peacetime, no more than five CONES were allowed
in a train, marshalled at the rear; during both wars,
these regulations were relaxed, and up to sixty
gunpowder vans could make up a train on govern-

ment explosives traffic. The official capacity
limitations were also altered, allowing 16,000lb
of explosives to be carried rather than 10,000lb;
if sheeted OPENS were used, 4,000lb could be
carried instead of 2,240lb.

GWR CONES at the turn of the century were
essentially the same as iron MINKS (cf Z1 and
V6), except that they had metal doors with flush
exterior plates, lacked end vents and were
marginally wider. They had a special black livery
with a large red 'X' painted right across the doors,
and the letters 'GPV' (gunpowder van) painted in
red on the sides and ends.

Z2 reflected constructional details about 1914,
the body having square corners with angle
reinforcement typical of iron loco coal wagons
(round-cornered metal wagons, with a few ex-
ceptions such as N22 and CC7, had ceased being
manufactured). The body style was continued in

Above : *Z2 gunpowder van, photographed when built 1914 showing lead interior lining, special 'magazine' boots and 'Rules to be observed in loading or unloading explosives', amongst which are 'Explosives must only be loaded or unloaded between sunrise and sunset' and 'This van when loaded or empty, must not be labelled so as to travel via the Severn Tunnel'. Square-cornered van. Iron u/f (Hartshill Iron Company channel), GWR details, V-axleguards, DC III brake.*

Left : *Z1 CONE built and photographed 1904. Essentially iron MINK (round corners) DC I brake.*

Below : *For Z2, u/f as O15 non-vacuum. For Z4, body extended to 16ft 6in oh, u/f as P18 (no steps).*

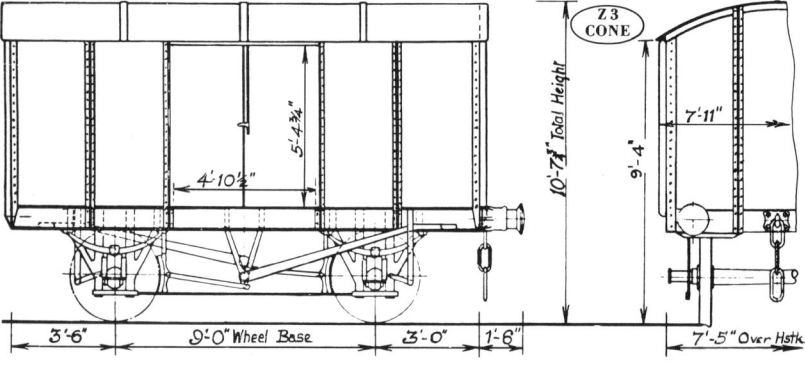

Z3 and Z4, differences arising from the contemporary underframe designs: Z2 (1914) had GWR self-contained buffers, DC III brake and 'wrap-under' side stanchions; Z3 (1924) had GWR self-contained buffing and drawgear, Morton brake and 'wrap-under' stanchions; and Z4 (1939) had RCH self-contained buffing and drawgear, Morton brake and 'straight-down' stanchions. Z4 was 6in longer than the others, using the N30 underframe, but rated at 7 tons (cf Z1). As far as the overall development of vans is concerned, the post grouping gunpowder vans retained the old 9ft wheelbase (RWA Fig 102). Note the oversize

wooden block for the label clip (sparks!).

In 1929, various gunpowder vans that had been converted from iron MINKS in the emergencies of World War I, were rebuilt for ordinary merchandise traffic (retaining no end vents however).

A lot was issued in 1937 for new CONES, but it was cancelled and various old iron MINKS were reconstructed instead. Of the 60 thus ordered, 50 were the same vehicles just described and ten extra were taken from the fleet of iron vans. Many of these were done at the behest of the Southern Railway, where they were sent on loan, lettered of course in SR style (RWA Fig 103).

CHAPTER 25
AA – BRAKE VANS
(Toads)

It has been said that 'once you have seen one GWR brake van, you have seen them all'. Certainly the basic single verandah design was used from the 1880s into nationalisation, but 23 diagrams were issued in the AA-index, and there were other variations omitted!

Most modern brake vans come from one of two groups: 20ft over headstocks, 13ft wheelbase, with a 6ft 6in verandah, built in the period 1888 to just before World War I, or 24ft over headstocks, 16ft wheelbase, with an 8ft 6in verandah, built thereafter into nationalisation. Before 1888, 18ft outside-framed brake vans were built. As with other series of vehicles, new diagrams were issued for changes in buffing, drawgear and other

See text for detailed differences between
AA1-4/6/9/12/14.

details with identical bodies. The tonnage of vans gradually increased over the years from 10 or 12 tons, to 16 and 20 tons; there were many 25-ton vans built at the turn of the century, but 20 tons later became standard. Channel underframes 1ft deep incorporated a hollow chassis which allowed scrap iron to be built into the vehicles for ballast weight. To improve braking adhesion, sanding gear was provided having a sand box at each corner of the chassis, two boxes on the verandah and two inside the van (where they often fitted into the seats and locker arrangement). Wet sand was originally used, but after World War I dried (locomotive) sand was adopted. Wet sanding vans had pipes which dropped sand straight down to the rails, ahead of the wheels, whereas with the free flowing dry sand, the pipes curved into the feet of the wheels. Two clasp shoes per wheel were customary on brake vans from early days. After the introduction of

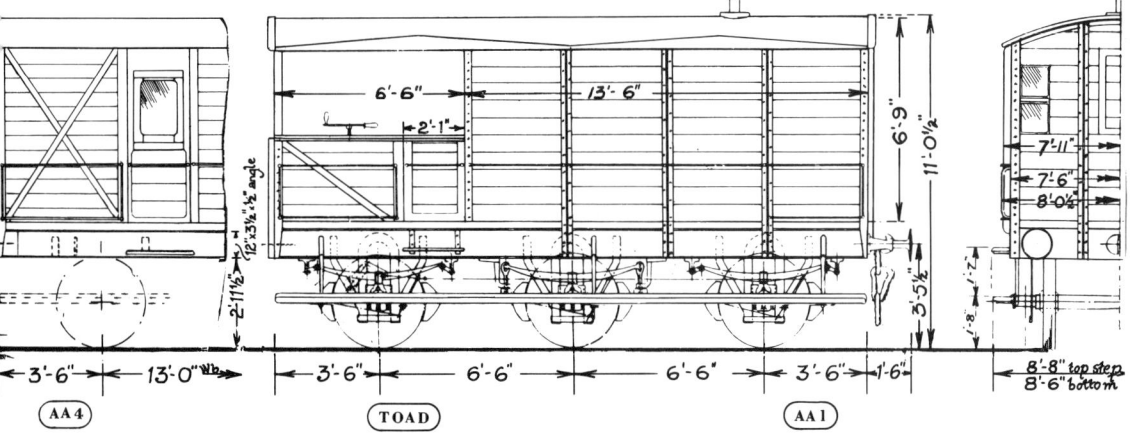

AA4   TOAD   AA1

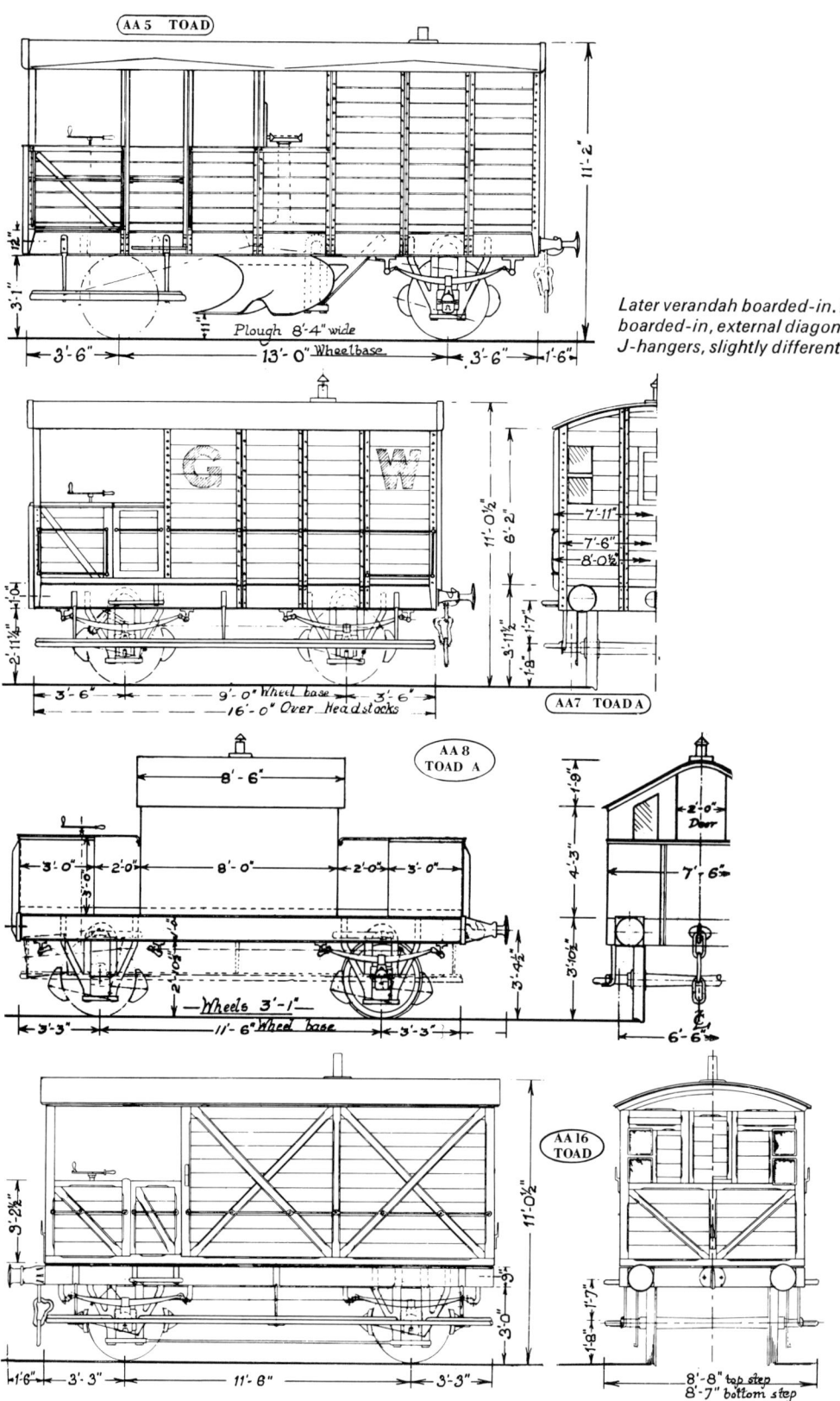

Later verandah boarded-in. For AA10, boarded-in, external diagonals, spring J-hangers, slightly different plough.

fitted trains, many vans were fitted with vacuum pipes (although not vacuum cylinders), the guard having a brake valve in the van; some vans were later built with a vacuum cylinder as well.

The oldest 20ft over headstocks design was AA3 (RWA Fig 97). More than 500 of these were built up to 1901, the earliest being 13-ton vans, later ones being 16 and 20 tons. A 25-ton version with the same body was produced during 1903-10, these heavy four-wheel vans being given the diagram AA2. The seemingly superfluous 'four-wheel' was used in the description since sixty-two 24-ton six-wheel vans, with 12 brakeblocks, had been built in 1901-2, again with the same body as AA3 (RWA Fig 88) as built; Fig 90-2 later including changed handrails. When the index was drawn up, these vehicles became AA1. From their home depots, in which Aberdare, Severn Tunnel Junction and Stoke Gifford figured prominently, they were used on the heavy coal trains. Actually there were minor differences between the bodies of AA3 and AA1/2; on the latter a seat was extended across the verandah between the sand boxes, and there was a similar seat inside the van.

Also 3ft 2in (rather than 3ft 1in) wheels appeared, which size remained standard thereafter.

The original list contained AA1-8. AA7 were twelve short wheelbase vans built in 1898 for working the Metropolitan Railway and stationed at Acton (RW p.121). Their 9ft wheelbase, 16ft over headstocks underframe was the same as contemporary four-plankers for example, but with 1ft channel, and the body was essentially a shrunken version of AA3 – the 5ft 3in verandah was the same proportion of 16ft, as 6ft 6in was of 20ft. These 12-ton vans must have been amongst the first to be fitted, in connection with the Smithfield meat trains.

AA8 were four special vans built in 1889-90 for the restricted clearances of the Pontnewynydd branch near Blaenavon, Newport Division. Of iron construction, they were double roofless verandah types (cf wooden brake vans of the 1860s), with a low central compartment (RWA Fig 82). Their underframes were similar to some very old vans (see AA16 later). Two iron brake vans for regular use had been ordered under osL388 in 1887 (assigned numbers 8774, 18ft over headstocks, and

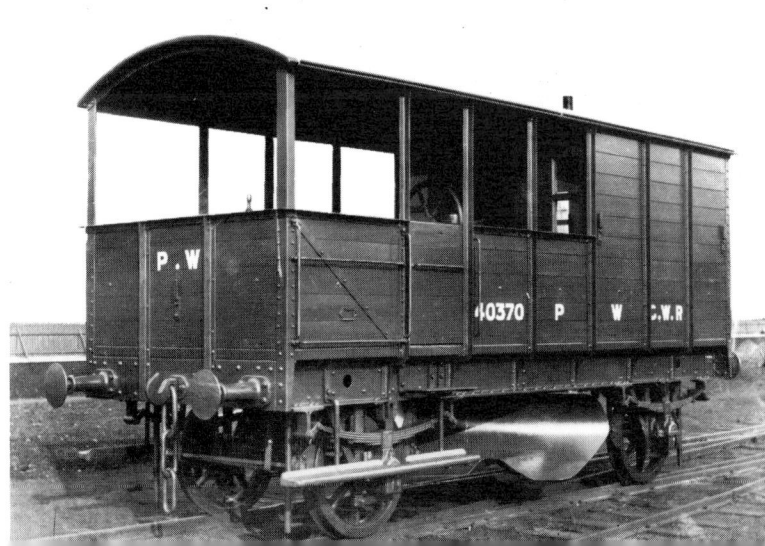

AA6 permanent way brakevan. Photographed as built 1890. Note track! Split 'stable-door'. No tare weight marked, was about 13 ton 10 cwt. Underframe like AA5, including metal footstep, but grease boxes.

Severn Tunnel brakevan. Built 1888 as open verandah 16 ton AA3 type, later closed in with drop light door. Photographed 1922. T-spring hangers.

AA15 20 ton 24ft over headstocks 'improved goods brakevan', introduced 1912 (AA11). Photographed when built 1919, 25in lettering. Steel sheeting on lower half of body. J-spring hangers. 8 by 4½in OK boxes. No tiebar between axleguards. Self-contained buffers. Commode rails attached to stanchions via plates, not by flattened ends as before. Chimney to one side; earlier central.

Goods Brake Van built and photographed circa 1888. Standard design, 18ft oh, before introduction of AA3. No tare load given. J-hangers, grease boxes, 'level' camber springs (cf G23 MAYFLY u/f). Note wagon-type brake gear on both sides, blocks appear to be reversible (?). No sanding gear. Body packed off 9in channel u/f. Metal plate top step. Handrails not painted white. Central white stripe (?) Some of these vans were uprated to 12 tons at end of World War I and given AA16.

7862, 19ft 10in over headstocks). Only one was built, the subsequent history of which is not known, and like the iron MEX the idea was not pursued, save for these AA8; however two new Pontnewynydd vans were ordered in 1947 under L1603 to replace (with the same running numbers) 7590/1; they were eventually outshopped in 1949 under the equivalent Western Region L2096 and given BR numbers. They were roughly to the original design, square verandah corners replacing round (RW p.125).

AA6 were permanent way brake vans and AA5 were the same 13½-ton vehicles fitted with 'spreaders' (ploughs for the hopper ballast trains); both were still coded TOAD. Again, the bodies were AA3, except that the verandah was boarded in (with windows to the front), and a split 'stable door' arrangement was provided; they doubled as mess and tool vans. Fifty-four vans to the AA6 diagram were built in the last decade of the nineteenth century, and 22 plough vans after 1893. Although built at the same time, there were four stanchions on the side of AA5, whereas AA6 had three as on all other types of goods brake van. As built, AA5 had a very long verandah so that the vertical handwheel which controlled the plough was out in the open. Since the screw mechanism was more or less in the middle of the underframe, the maximum closed compartment available was only some 8ft 3in long. Bounding the cabin with the conventional stanchion positions would have given only a 6ft 9in van. Hence a greater number of evenly spaced stanchions was used. Apparently it proved unnecessary for the plough operator to stand in the open, for ultimately the bodies were boxed-in; it is in this form that they appeared in the index.

Finally, AA4 relates to a brake van built specially for working the Severn Tunnel. It could not have been pleasant amongst all the smoke and fumes, and six years (!) after the opening 35953 was built with closed ends. The only difference from AA6 was that the door was a single carriage-type door with a droplight.

During 1905-8, five new permanent-way brake-vans were added to the list. Heavy 25-ton vehicles, with additional X-bracing on the AA6 body, they were given AA9. Again in the 25-ton vogue, AA10 of 1909 was another plough van, additional diagonal bracing being provided. Although probably built with closed ends, the same stanchions as AA5 were used.

L707 of 1912 introduced the 'improved goods brake van', 56483, which had an 8ft 6in verandah and 15ft 6in cabin to give 24ft over headstocks;

vacuum fittings and screw couplings were provided: 13-, 16-, and 20-ton versions were built all to this diagram (AA11), which like AA10 had extra diagonal bracing. A detail difference on this and later vans was that a lever was hung from the roof of the verandah near the compartment door, with rodding connected to the rear sand boxes (it also acted as a handrail inside the van), so that the guard did not have to run between sand boxes; he was also provided with a sloping-top desk. Fig 89 in RWA shows the interior of such a van. The stove chimney stack was still straight through the top on AA11 but it went to the side on AA13. After the 20-ton vans of L757 in 1913, steel sheeting covered the verandah sides, and the spring hangers changed to underslung. On the new diagram thus issued, AA13, no diagonal bracing was used. Steel sheeting appeared along the bottom of the whole van on AA12, which were some 20ft over headstocks ex-permanent way vans converted for ordinary goods use. Similar sheeting was intended to be used on twenty new heavy permanent way vans ordered in 1914 to diagram AA14. Allotted 60771-90, only one was built because of the war; it was numbered 14678 and stationed at Maindee, Newport. The order for AA14, like some of P14, was closed on 13 March 1928.

Immediately after the war, more 20-ton vans to the 'improved' design were built. Since they had GWR self-contained buffing and drawgear the new diagram AA15 was issued. Thereafter the basic design changed only little by little into nationalisation: AA18 reverted to old-type spring shackles and had new running board hangers. Gone were the elegant rods of before! The body was AA15, but with single panes of glass in the windows (a feature that had appeared on two closed-ended AA15 vans assigned for Severn Tunnel working, which were given AA17); AA19 of 1927 incorporated as much RCH detail as possible (buffing and drawgear, wheels, axles and axleguards) and had a structural steel underframe; AA20 provided inside lamp brackets, lamp-irons and a padded armchair for the guard (a Severn Tunnel Junction version of this design was issued under AA22, the low side sheeting now covered the bottom of the door); a change occurred here too with the handrails. AA21 (Vol.1, p.75) were 100 vacuum cylinder vehicles (not included in actual stock) financed by the government at the outset of World War II, and AA23 of 1942 were non-vacuum versions of AA21 (Vol.1, p.76), an added feature was a lever for the front sanders, which had also been seen on AA22. A new chimney also appeared on this design which lasted to BR days.

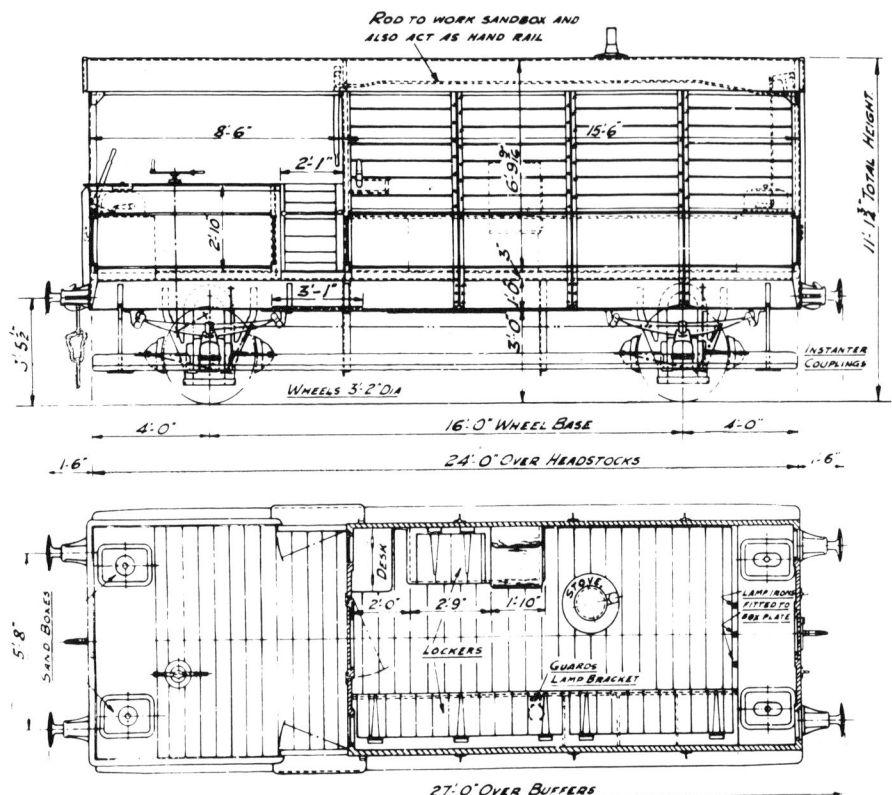

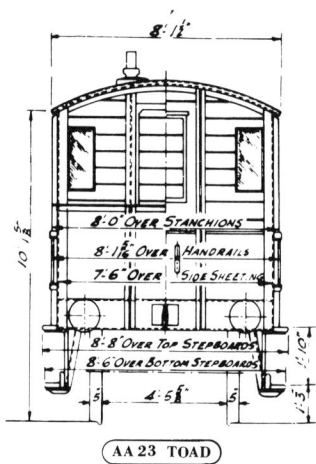

AA 23 TOAD

See text for detailed differences between A11/13/15/17-22.

Many vans were altered from their original appearance when repaired. Hence steel casing appeared on early vans, such as AA1; the original flat cross-bracing on the boarded-in verandahs of AA5/6 disappeared; extra diagonals appeared on the sides of other vans, both V-like as on AA11 and A-like, for which no diagram was issued rainstrips differed widely between vans. Incidentally, the AA11 index was abandoned and all the vans lumped into AA13 (cf M3/4).

Towards the end of World War I, there was a shortage of light brake vans. Consequently various old 10-ton vans with outside-framed 18ft over headstocks bodies (which did not have a diagram number) were reconstructed (to include GWR self-contained buffers) to 12-ton vehicles and given AA16. A few survived until 1951; others were converted to weighing machine vans for Messrs Pooley & Son, to tool vans for the signal department, or to tunnel inspection vans. The original vehicles were the standard brake vans before the introduction of AA3. Some were built with closed ends as ballast train vans (osL364 in 1886) and it is probable that some were also equipped with ploughs (cf osL326 in 1884 for GWR/LNWR joint lines), while at another extreme, the parents of the design were first built in the 1870s with roofless verandahs. Some typical running numbers for the AA16 type in their original guise were: 3259/3922/8646/8782 (later converted to G23 MAYFLY); 17501-700/25-826 etc; 35961-36000.

CHAPTER 26
BB – STORES VANS

Only one diagram made up this group. The 18ft wooden body design originated in 1894 in part of L9, the rest of that lot being 16ft iron MINKS. Fitted to run in passenger trains, the wagons had Armstrong vacuum fittings on a 12ft wheelbase, and were among the few GWR vans to have sliding doors. The eleven original vehicles had unusual open-spoke coach-size wheels. Two 8-ton vans were built in 1902 for the Plymouth division (marked 'Return to Taunton Engineering Department').

Later on, stores ENPARTS were carried in converted vehicles (cf V9) for which no separate diagrams were issued. Likewise some of the early iron MINKS were assigned to the loco department to work between Swindon and Stafford Road, but these also were never put into the BB lists.

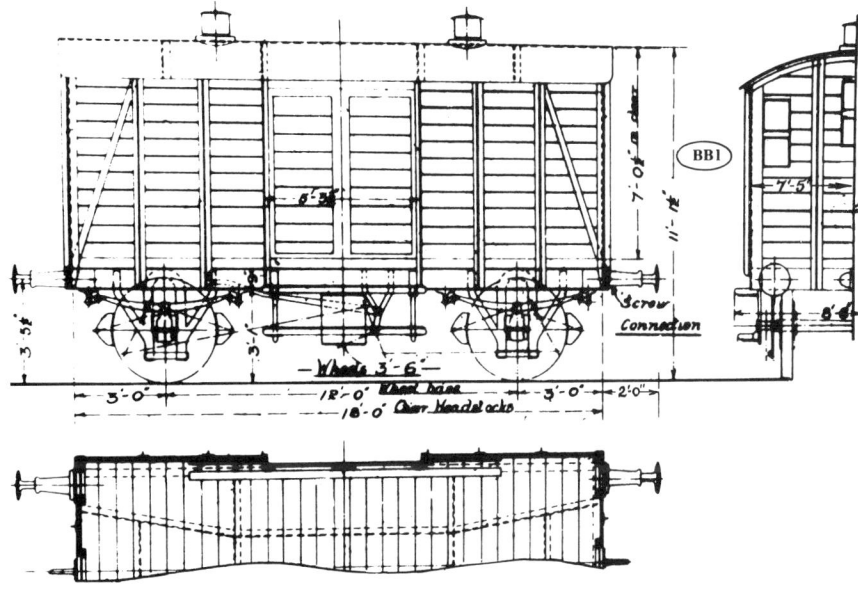

BB1 stores van photographed when built 1894. Note track! Sliding doors, drop pin chain no loop. Iron roof, oil lamps. Armstrong vacuum u/f (long lever this side, toothed rack), clasp shoes. Release lever star under door. X for oilboxes 1ft 8½in buffers, safety chains T-hangers. Unusual open-spoke 3ft 6in wheels.

CHAPTER 27
CC – WORKSHOP AND TOOL VANS, AND CRANE TESTING VANS

This group included vans for both the signal and engineering departments and for the use of Pooley & Son (which was responsible for weighbridge maintenance). In the early days, iron tool vans were popular; 18ft over headstocks, 11ft wheelbase adaptations of the iron MINK design became CC1/2 (with 4ft 6in springs slung on T-hangers), and a six-wheel 24ft 6in design built between 1902-12 became CC3. All these vehicles, and most others in the group, were characterised by oil

lamps, windows in the sides and ends, and skylights in the roof; the vans were equipped with work benches and lockers. CC2 (for Pooley) were the same type as CC1, with the addition of a pocket under the wagon, Vol.1, p.21. The crane testing vans CC4 had also essentially the same body as CC1, with different window arrangements, but the underframe had 3ft 7in passenger wheels, Armstrong vacuum and 2ft buffers (cf K2 match trucks).

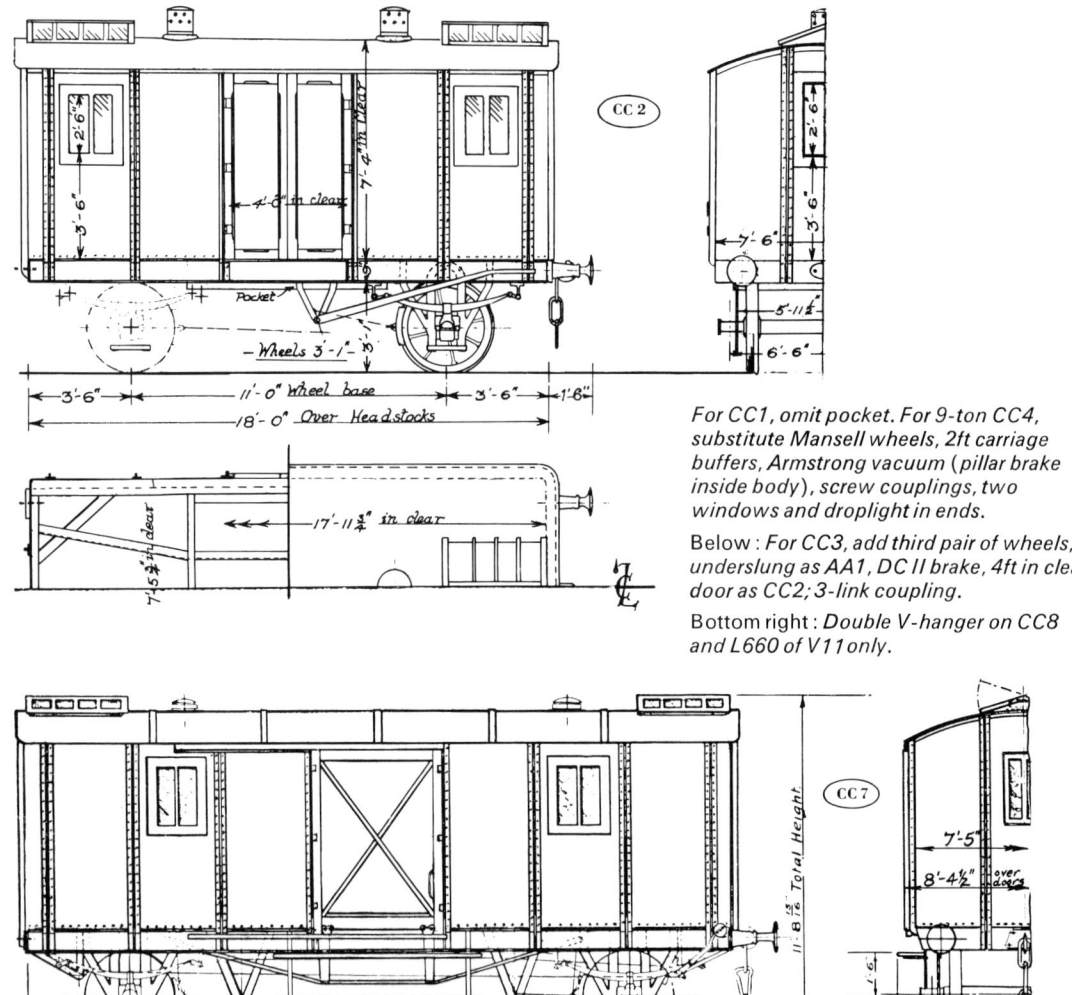

For CC1, omit pocket. For 9-ton CC4, substitute Mansell wheels, 2ft carriage buffers, Armstrong vacuum (pillar brake inside body), screw couplings, two windows and droplight in ends.

Below: For CC3, add third pair of wheels, underslung as AA1, DC II brake, 4ft in clear door as CC2; 3-link coupling.

Bottom right: Double V-hanger on CC8 and L660 of V11 only.

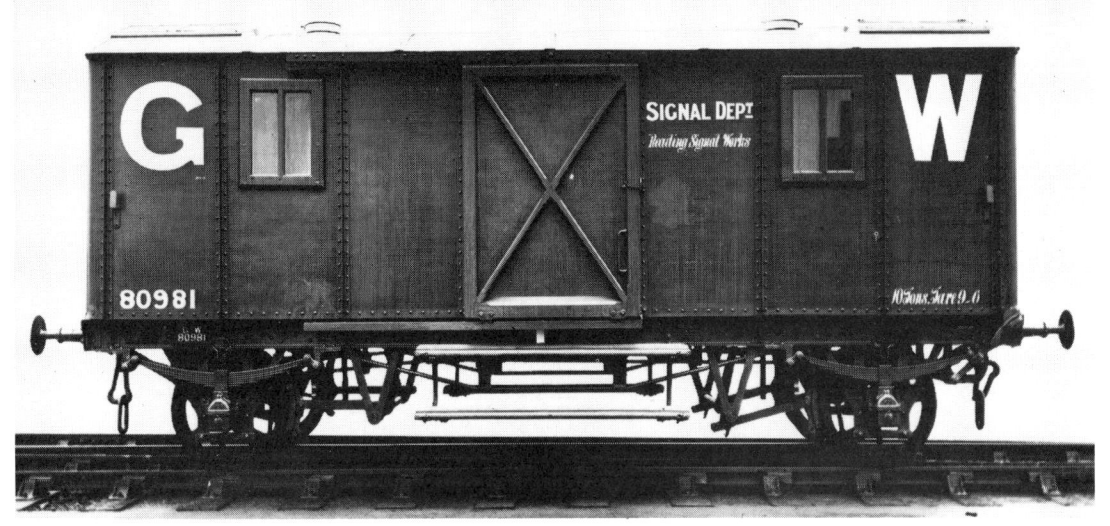

Top: *CC7 tool van. Built and photographed 1913. Stretched out CC2. Toplights, lower vents on lamps. Sliding door. Fixed rod trussing, Γ-hangers, DC III brakes (kinked handle—not painted white).*

Above: *CC8 tool and packing van. Photographed when built 1913 (brown livery) V11-type 28ft 6in body with addition of end door, skylights, and oil lamps. Underslung J-hangers. 8 by 4in OK boxes. DC III brakes, white stripe.*

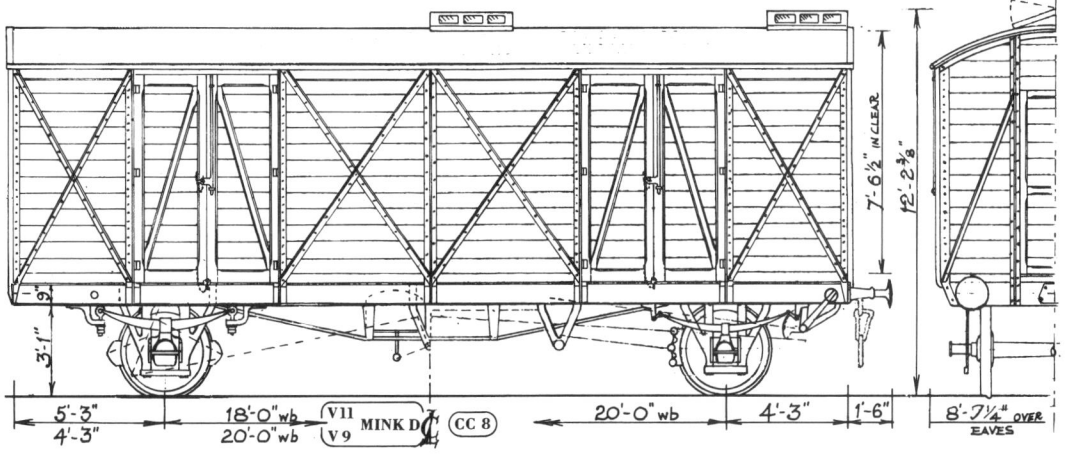

CC2 iron workshop van for use of Henry Pooley & Son,
contractors to GWR for weighing machine repairs.
Built and photographed 1899. Taller and longer than
iron MINK, narrow 4ft doors. Oil lamps, skylights in
iron roof (bands). T-hangers.

Wooden bodies on the smaller vans appeared after about 1910. Thus CC5 were four V12-type vans, with narrow doors like CC2, for Pooleys. Windows caused slightly unusual bracing. Two were built before World War I, two after. These latter were noted in the lots to have 'spring rod buffers and drawgear', since self-contained details would be normal at that time. CC6 and CC8 were 'one-off' wooden tool and packing vans for Swindon and Wolverhampton works respectively. A V16-type MINK (conventional door width) was given four lights and correspondingly altered bracing for CC6. The 28ft 6in V11 MINK D design was likewise modified for CC8; end doors, and three skylights were used.

Finally, the iron body of CC3 was put on a four- instead of six-wheel underframe just before World War I to become CC7. A sliding door replaced the pair of swing doors, and later lots of the early 1920s included a 10cwt jib crane inside the door (RWA Fig 261). These vehicles were the last round-cornered GWR vans to be built.

As in the BB group, it became the practice to convert old vehicles for workshop vans rather than build new. Thus MINKS and AA16-type brake vans were rebuilt, and in December 1922 an ex-Cambrian Railways meat van was converted for Pooley's use in the Newtown, Montgomeryshire, District. Likewise L1514 in 1945 related to the conversion of a brake third 'to replace G-shop tool van, 43968'; this van was built as a CC4 in 1900.

Some original assignments were CC2 16910 Newport; CC3 14988 Teignmouth; 14983 West-bourne Park; 14932 Neath; 14991 & 80999 Wolverhampton; 14989 St Blazey; 14987 Taunton and 80989 Shrewsbury.

CHAPTER 28
DD – TANK WAGONS
(including Cordons)

The GWR, along with most other railway companies, owned relatively few tank wagons. Nevertheless, a few locomotive department water tanks, and oil and creosote tanks were built for the company's use.

The earliest tanks were not cylindrical, but rectangular. A 2000 gallon design dating from the early 1890s (DD2) was essentially the iron coal wagon N6 with a roof and iron floor, plus a manhole and valves etc. DD1 was a lengthened version of this, 20ft over headstocks, 13ft wheelbase which held 3000 gallons, (RWA Fig 78 in 1904 painting trials with large lettering; cf same running number used by DD4 gas tank wagons). A 19ft over headstocks, 12ft wheelbase cylindrical design appeared in 1895 (DD3). It is interesting to note that the 7ft

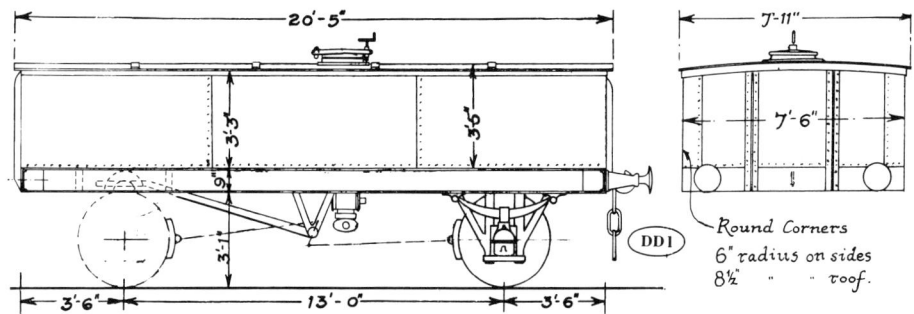

Above : *For DD2, 16ft oh, 9ft wb, tank 3ft high side.*

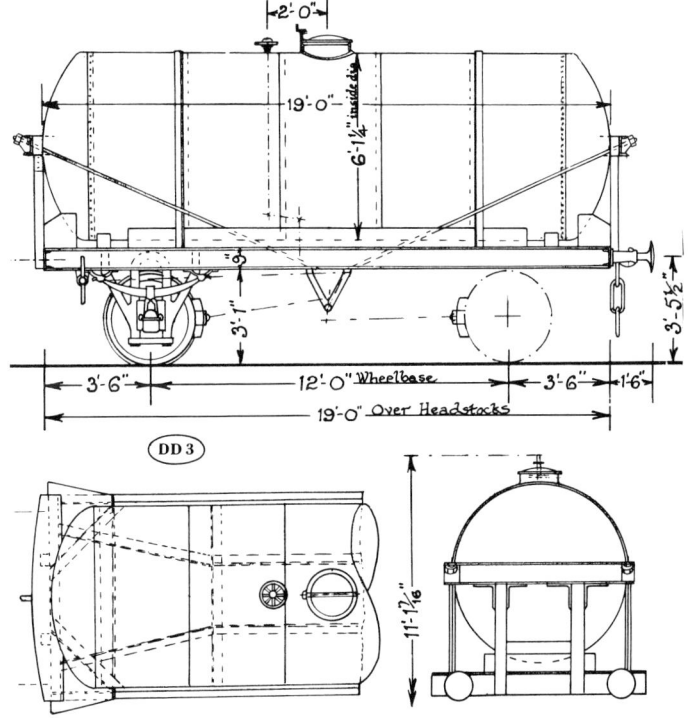

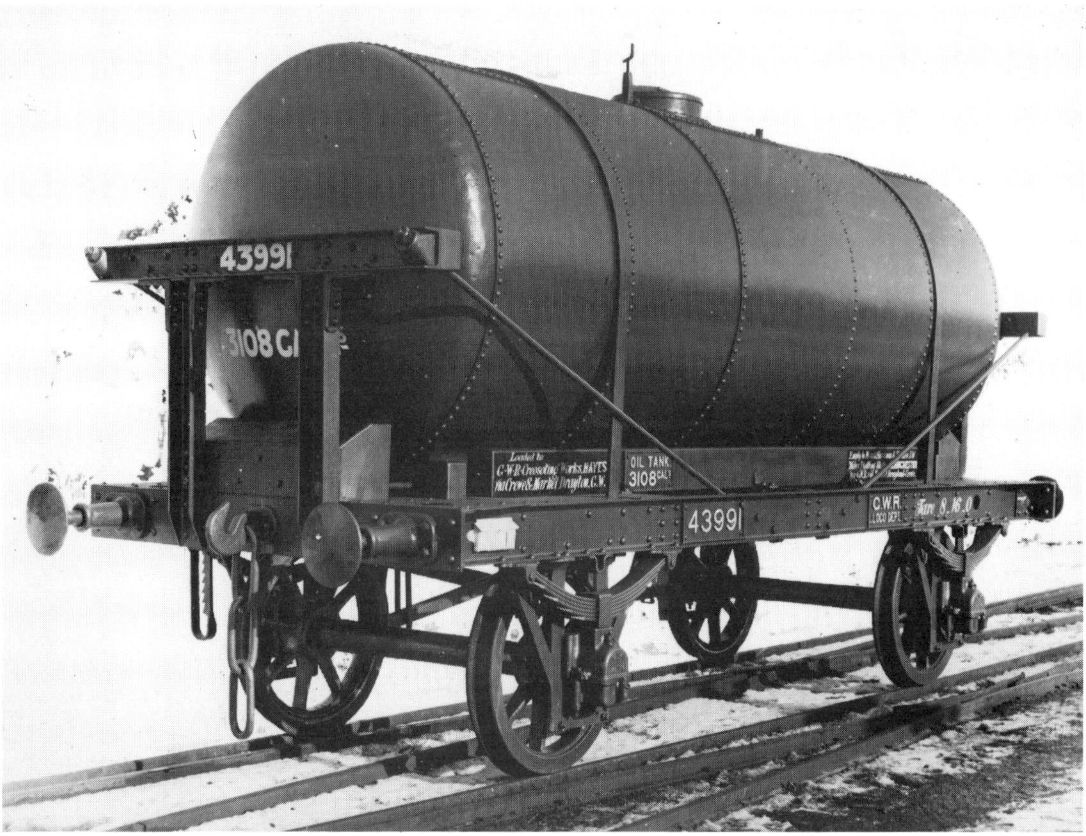
43991
OIL TANK
3108
43991
G.W.R.
LOCO DEPT.

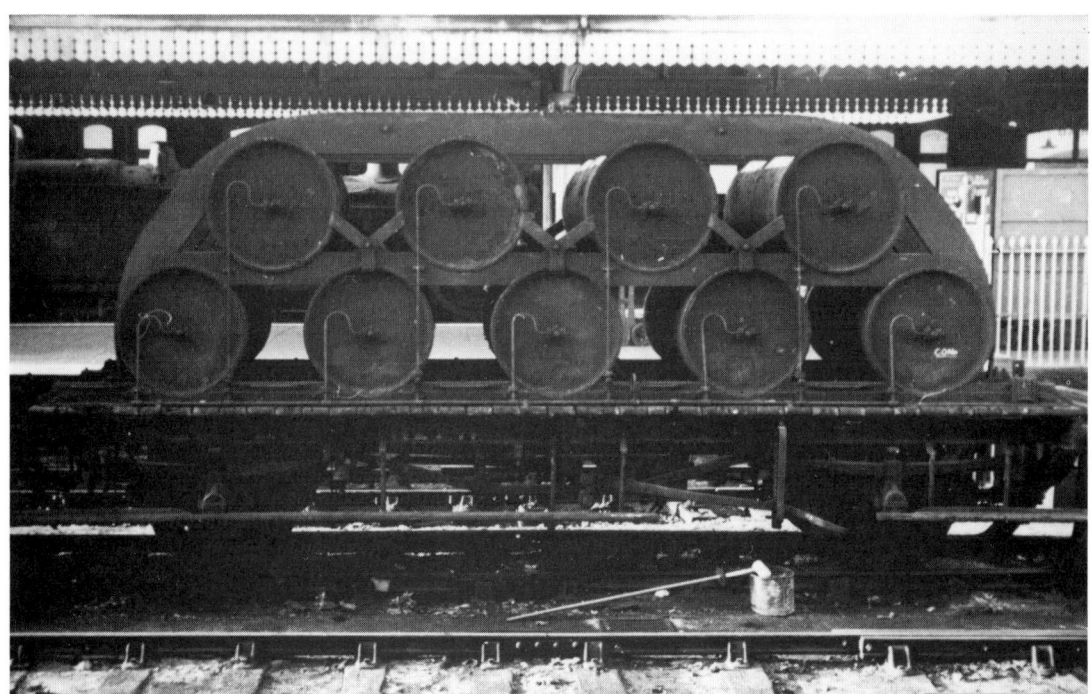

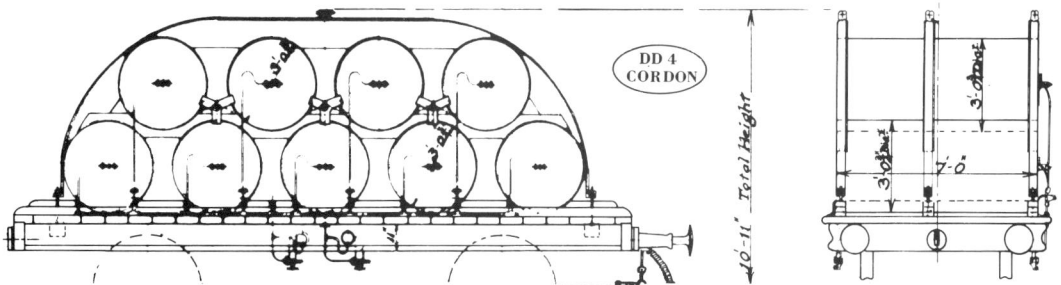

DD 4
CORDON

Left: *DD3 built and photographed 1895. From the
first lot, frames built by wagon dept, tanks by
locomotive dept. One-sided lever brake, toothed rack.
Cast plates and painted ends.*

Below left: *DD4 CORDON travelling oil gas tank wagon.
Nine transverse tanks yoked onto second-hand
carriage underframe (given angle trussing). 4ft 6in
springs. J-hangers. Long lever brake (interrupted
step boards) Morton cam this side.*

(K. Williams)

Below: *DD5 CORDON. Two longitudinal cylinders.
Old carriage underframe, Armstrong vacuum, disc
wheels. Others had DC quadrant levers, or pillar brake
handles on end platforms (see also RWA Figs 123/5).*

(K. Williams)

diameter by 19ft domed tanks of $\frac{3}{8}$in plate (holding
the strange number of 3108 gallons) were built by
the locomotive department, whereas the frames
were built by the wagon department. Forty one of
these vehicles were constructed by 1911 (the
earliest of which had the Thomas brake) and one
further was ordered in 1914 (L788) but never built.
They were used variously for carrying ammoniacal
liquor, gas oil, creosote, lubricating oil, and also
water. Various tank top fittings thereby appeared
(RWA Figs 126/8/30). For example, 43991 built
in 1895 carried creosote between Messrs Harden
and Holden, Manchester and Hayes (via Market
Drayton and Crewe). Other gas liquor tanks were
mounted on old MACAW underframes formerly
numbered in the 3xxxx series (e.g. 43915-20 in
1910, cf also RWA Fig 120). A tank wagon was
built for the government during World War I for
which DD6 was issued. The diagram fell from use

after the war, and was not re-used until new 6-wheel water wagons (adapted from contemporary milk tank wagons) appeared just before nationalisation (cf DD7, RWA Fig 131 which had a GWR building plate).

When gas lighting became common, tank wagons for supplying stations became necessary. A number of different designs for travelling gas tanks appeared (RWA Fig 121 et seq), collectively coded CORDON, but diagrams were issued for only two types. DD4 were the well-known transverse tank design. Secondhand carriage underframes (some wooden) were used to mount nine 3ft diameter by 7ft tanks across the wagon, four on five, the total capacity of which was 403 cu ft; three strapped wooden yokes held everything in place. DD5 was a twin-tank design, two 4ft diameter by 15ft cylinders being mounted side-by-side on old

18ft 6in over headstocks chassis. An odd feature on some of DD5 was the pillar brake, like that on U1 and the early Birkenhead shunting trucks. Detail differences occurred between wagons, such as clasp brakeshoes, DC brakes etc, depending upon the original underframe and time of conversion. Number 72 was sold to the War Office in 1917, and although a replacement was supposed to have been built, the lot was cancelled. Some of the wooden framed vehicles were later rebuilt with iron frames (e.g. CORDON 52 in 1941).

For reference the gas system operated at 140psi, and a typical incandescent burner would use about 7 cu ft/hour. CORDONS were painted black with white lettering.

In addition to those listed in Chapter 3, there were other CORDONS built on carriage lots, all numbered between 1 and 100.

CHAPTER 29
EE – FLAT WAGONS FOR DEMOUNTABLE TANKS

Tank wagons for liquids in bulk were admirably suited for transport between rail-connected premises, but where such facilities did not exist it was the practice to forward the liquids in drums, casks or cans, or else for tank wagons to be decanted at

the destination station for delivery. Just before World War II the railway companies put into service road-rail tanks for the conveyance of large consignments of liquids from door to door for firms with no sidings. They were of two types, tanks

EE2 converted and photographed 1947. Container flat wagon 4 by 500 gal welded demountable tanks for ICI paints and varnish. Underframe from 6-wheel SIPHON no.1931.

FF1 Trestle plate wagon. Photographed when converted from O37 HIGH in 1942. Six planks homegrown oak, 9in corner plates, exposed floorboards, gusset attachment of diagonals to base of 'straight-down' L-stanchions. Sacktruck (tapered foot) door. RCH long rib buffers (top ribs also), RCH u/f fittings. Morton brake, 2 blocks, toothed rack.

mounted on road trailers which were hauled at terminal points by tractor, and demountable non-wheeled tanks which were transferred from road vehicle to rail wagon and vice versa by crane. The former type were incorporated into the GWR carriage index but the latter began the EE group in the wagon index. As the name implies, these wagons were for 'container tanks', and indeed many were converted from CONFLATS with special extra scotches. Order 5886 of 1938 fitted up

36520 (H7) to carry a single 1300gal tank for Unilever's traffic in oils and fats. Others from H7, and the ex-milk tank underframe H8 followed soon after the war and became EE1. EE2/3 had four 500gal tanks each for ICI paints and varnish. Here the underframes of both old six-wheel and bogie SIPHONS were used. Some liquids required a high temperature to be maintained during transit, which was achieved by electrical heating generated by a dynamo on the wagon.

CHAPTER 30
FF – TRESTLE PLATE WAGONS

In July 1942, ten open wagons of L1360 (O37) were equipped with a trestle the length of the wagon, for which the FF index was used. Sent to Guest Keen & Baldwins, Port Talbot, where they

joined the CORALS, they were downrated to 9 tons for the conveyance of steel plate. Ultimately they returned to merchandise goods traffic for general service.

BIBLIOGRAPHY

The principal references were unpublished books and documents such as:
GWR Lot Lists
GWR Diagram Index for Wagons for various years
GWR Telegraph Message Code Book for various years
GWR Circulars
Swindon wagon notebooks
Special Wagon Book for various years
Mr R. Woodfin's records
The chief published references were:
Books:
Bell, R., *The History of British Railways During the War 1939-45*, Railway Gazette
Essery, R. J., Rowland, D. P. and Steel, W. O., *British Goods Wagons*, David & Charles, Newton Abbot, 1971 (referred to as BGW)
Holcroft, H., *Great Western Locomotive Practice, 1837-1947*, LPCo, 1957
McDermott, E. T. and Nock, O. S., *History of the Great Western Railway*, republished by Ian Allan, Shepperton, 1967

Mountford, E. R., *Caerphilly Works, 1901-1964* Roundhouse Books, 1965
Pratt, E. A., *British Railways in the Great War*, Vols. I & II, Selwyn & Blount, 1919
Russell, J. H., *A Pictorial Record of Great Western Wagons*, Oxford Publishing Co, Oxford 1971 (referred to as RW)
Russell, J. H., *A Pictorial Record of Great Western Coaches*, Part 1, Oxford Publishing Co, Oxford 1972 (referred to as RC)
Russell, J. H., *Great Western Wagons Appendix*, Oxford Publishing Co, Oxford 1974 (referred to as RWA)
Stone, S., *Railway Carriages & Wagons: Their Design and Construction*, Railway Engineer, 1911
Periodicals:
GWR (London) Lecture & Debating Society Proceedings
Great Western Echo
GWR Magazine
The Locomotive
Railway Engineer
Railway Gazette
Railway Magazine

ADDENDUM TO VOLUME 1

Following publication of Volume 1 a number of points in Chapter 3 The Index of Wagon Diagrams, table 7 have been clarified or amended and are summarised here under the headings of the appropriate diagram number.

B B8 Nos 32219/30 should be 32219-30

C C8 osL 373 No 36939 should be 36949
C10 Indistinct number in L198 is 33988
C19 Numbers are 41952-4

G G32 Add Nos 116989-7000
G43 Add No 65789, total 100 wagons, not 109

J J7 L595 numbers should read 14430
J11 L553 numbers should read 70797-806
J21 L873 should read L878
J30 L1332 should read L1382

M M4 Nos 41046-55/91-41000 should read 41046-55/96-41100

N N27 Add Nos 33457-9/919-37
N34 L1480 Add Nos 63129-48

O O9 Nos 81001-880 should read 81001-380
O18 No 9420 should read 94201, and add Nos 98651-9150
O32 Nos 13337-696 should read 134337-135896
O37 L1370 should be L1379
O38 Nos /517-832 should read 35170-832

P P7 Final Lot No /540 should read /54
P15 Lot Nos should read 1307/16/19/28

T T7 L765 No 14423 should read 14457
T8 L792 add No 14680

V V12 Nos 73/310/8445 should read 73/310/8/445
V14 Nos 93590-3839/5674-823 should read 93590-789/5674-823

AA AA4 Against osL 648, the lot list said 'as (os)L 432'; osL 432 of 1888 was the first 'modern' brakevan design (i.e. not externally wooden framed) and has been attributed to AA3. If in fact it was a Severn Tunnel van, it would be closer to the date of the tunnel being opened, and also might explain why the design was changed.
AA12 L799 was split between AA12/13 as
AA13 follows:
AA12, 22½ tons, 17787-91/93-821;
AA13, 20 tons, 17823-38.
AA23 L1451 Nos 35928-52/4/70/4- should read 35928-52/4-70/4-

INDEX

This index lists wagon types either by code name or by wagon diagram book index letter, and where dealing with the overall class the total chapter extent is shown. Individual references to types in chapters outside their main description are shown separately. This index also shows which types are illustrated by drawings (in italic type) or photographs (roman type) within the chapter extents (or elsewhere). Vehicle names and telegraphic codes not shown here are in the contents list on page 3. A full list of vehicle telegraphic codes and respective diagram numbers are given in Volume 1, chapter 2.

Authors' concluding note

We would like to repeat here that while we hope that we have prevented errors from slipping through this work, we shall be glad of any comments and corrections from fellow students of Great Western history which can be sent to us through David & Charles. We trust that these books will promote the publication of both the mass of hitherto unseen information and that which must still as yet be undiscovered, so that we can all continue our study of the fascinating Great Western Railway.